Lecture Notes in Electrical Engineering 1483

The book series *Lecture Notes in Electrical Engineering* (LNEE) publishes the latest developments in Electrical Engineering—quickly, informally and in high quality. While original research reported in proceedings and monographs has traditionally formed the core of LNEE, we also encourage authors to submit books devoted to supporting student education and professional training in the various fields and applications areas of electrical engineering. The series cover classical and emerging topics concerning:

- Communication Engineering, Information Theory and Networks
- Electronics Engineering and Microelectronics
- Signal, Image and Speech Processing
- Wireless and Mobile Communication
- Circuits and Systems
- Energy Systems, Power Electronics and Electrical Machines
- Electro-optical Engineering
- Instrumentation Engineering
- Avionics Engineering
- Control Systems
- Internet-of-Things and Cybersecurity
- Biomedical Devices, MEMS and NEMS

For general information about this book series, comments or suggestions, please contact leontina.dicecco@springer.com.

To submit a proposal or request further information, please contact the Publishing Editor in your country:

China

Jasmine Dou, Editor (jasmine.dou@springer.com)

India, Japan, Rest of Asia

Swati Meherishi, Editorial Director (Swati.Meherishi@springer.com)

Southeast Asia, Australia, New Zealand

Ramesh Nath Premnath, Editor (ramesh.premnath@springernature.com)

USA, Canada

Michael Luby, Senior Editor (michael.luby@springer.com)

All other Countries

Leontina Di Cecco, Senior Editor (leontina.dicecco@springer.com)

**** This series is indexed by EI Compendex and Scopus databases. ****

Ji Su Park · Jin Wang · Yi Pan · James J. Park
Editors

Advanced Future Information Technology

Proceedings of Future Tech 2025

Springer

Editors
Ji Su Park
Department of Computer Science
Jeonju University
Jeonju-Si, Korea (Republic of)

Jin Wang
Computer Sicence and Engineering
Hunan University
Xiangtan City, China

Yi Pan
Shenzhen Institute
Shenzhen, China

James J. Park
National University
Seoul, Korea (Republic of)

ISSN 1876-1100 ISSN 1876-1119 (electronic)
Lecture Notes in Electrical Engineering
ISBN 978-981-95-1998-9 ISBN 978-981-95-1999-6 (eBook)
https://doi.org/10.1007/978-981-95-1999-6

This Springer imprint is published by the registered company Springer Nature Singapore Pte Ltd.
The registered company address is: 152 Beach Road, #21-01/04 Gateway East, Singapore 189721, Singapore

Message from the FutureTech 2025 General Chairs

FutureTech 2025 is the 20th event of the series of international scientific conference. This conference takes place on April 2–26, 2025, in Zhangjiajie and Changsha, China. The aim of the FutureTech 2025 is to provide an international forum for scientific research in the technologies and application of information technology. FutureTech 2025 is the next edition of FutureTech 2024 (Chongqing, China), FutureTech 2023 (Phnom Penh, Cambodia), FutureTech 2022 (Jeju, Korea), FutureTech 2021 (Jeju, Korea), FutureTech 2020 (Jeju, Korea), FutureTech 2019 (Xian, China), FutureTech 2018 (Salerno, Italy), FutureTech 2017 (Seoul, Korea), FutureTech 2016 (Beijing, China), FutureTech 2015 (Hanoi, Vietnam), FutureTech 2014 (Zhangjiajie, China), FutureTech 2013 (Gwangju, Korea), FutureTech 2012 (Vancouver, Canada), FutureTech 2011 (Loutraki, Greece), FutureTech 2010 (Busan, Korea, May 2010) which was the next event in a series of highly successful the International Symposium on Ubiquitous Applications & Security Services (UASS-09, USA, Jan. 2009), previously held as UASS-08 (Okinawa, Japan, Mar. 2008), UASS-07 (Kuala Lumpur, Malaysia, August, 2007), and UASS-06 (Glasgow, Scotland, UK, May 2006).

The conference papers included in the proceedings cover the following topics: Hybrid Information Technology, High-Performance Computing, Cloud and Cluster Computing, Ubiquitous Networks and Wireless Communications, Digital Convergence, Multimedia Convergence, Intelligent and Pervasive Applications, Security and Trust Computing, IT Management and Service, Bioinformatics and Bio-inspired Computing, Database and Data Mining, Knowledge System and Intelligent Agent, Game and Graphics, and Human-centric Computing and Social Networks. Accepted and presented papers highlight new trends and challenges of future information technologies. We hope readers will find these results useful and inspiring for their future research.

We would like to express our sincere thanks to Steering Chair: James J. (Jong Hyuk) Park (SeoulTech, Korea), Doo-Soon Park (Soonchunhyang University, Korea), Jin Wang (Hunan University of Science and Technology, China). Our special thanks go to the Program Chairs: Ji Su Park (Jeonju University, Korea), Yan Li (Inha University, Korea), Ka Lok Man (Xi'an Jiaotong-Liverpool University, China), S. Vimal (National Engineering College, India), Le Anh Ngoc (Swinburne University of Technology, Vietnam), Xiaoliang Wang (Hunan University of Science and Technology, China), all Program Committee members, and all reviewers for their valuable efforts in the review process that helped us to guarantee the highest quality of the selected papers for the conference.

We cordially thank all the authors for their valuable contributions and the other participants of this conference. The conference would not have been possible without their support. Thanks are also due to the many experts who contributed to making the event a success.

Jungho Kang (Leading Chair)
Yang Xiao
Changhao Piao
Laurence T. Yang
Wei Liang
FutureTech 2025 General Chairs

Message from the FutureTech 2025 Program Chairs

Welcome to the 19th International Conference on Future Information Technology (FutureTech 2025), which will be held in Zhangjiajie and Changsha, China, on April 2–26, 2025. FutureTech 2025 will be the most comprehensive conference focused on the various aspects of information technologies. It will provide an opportunity for academic and industry professionals to discuss recent progress in the area of future information technologies. In addition, the conference will publish high-quality papers which are closely related to the various theories and practical applications in multimedia and ubiquitous engineering. Furthermore, we expect that the conference and its publications will be a trigger for further related research and technology improvements in these important subjects.

For FutureTech 2025, we received many paper submissions; after a rigorous peer review process, we accepted only articles with high quality for the FutureTech 2025 proceedings, published by the Springer. All submitted papers have undergone blind reviews by at least two reviewers from the technical program committee, which consists of leading researchers around the globe. Without their hard work, achieving such a high-quality proceeding would not have been possible. We take this opportunity to thank them for their great support and cooperation. Finally, we would like to thank all of you for your participation in our conference and also thank all the authors, reviewers, and organizing committee members. Thank you and enjoy the conference!

Ji Su Park
Yan Li
Ka Lok Man
S. Vimal
Le Anh Ngoc
Xiaoliang Wang
FutureTech 2025 Program Chairs

Organization

Honorary Chair

Young-Sik Jeong	Dongguk University, Korea

Steering Chairs

James J. Park	SeoulTech, Korea
Doo-soon Park	Soonchunhyang University, Korea
Jin Wang	Hunan University of Science and Technology, China

General Chairs

Jungho Kang	Baewha Women's University, Korea
Yang Xiao	The University of Alabama, USA
Changhao Piao	Chongqing University of Posts and Telecommunications, China
Laurence T. Yang	St. Francis Xavier University, Canada
Wei Liang	Hunan University of Science and Technology, China

Program Chairs

Ji Su Park	Jeonju University, Korea
Yan Li	Inha University, Korea
Ka Lok Man	Xi'an Jiaotong-Liverpool University, China
S. Vimal	National Engineering College, India
Le Anh Ngoc	Swinburne University of Technology, Vietnam
Xiaoliang Wang	Hunan University of Science and Technology, China

Local Chairs

Mingjie Liu	Chongqing University of Posts and Telecommunications, China
Ping Liu	Chongqing University of Posts and Telecommunications, China
Jianyong Yu	Hunan University of Science and Technology, China
Naixue Xiong	Hunan University of Science and Technology, China
Kuanching Li	Hunan University of Science and Technology, China
Shiming He	Changsha University of Science and Technology, China
Jiaohua Qin	Central South University of Forestry and Technology, China
Yuzhen Liu	Hunan University of Science and Technology, China
Wenti Huang	Hunan University of Science and Technology, China

International Advisory Committee

Victor Leung	University of British Columbia, Canada
Hsiao-Hwa Chen	National Cheng Kung University, Taiwan
C. S. Raghavendra	University of Southern California, USA
Philip S. Yu	University of Illinois at Chicago, USA
Hai Jin	Huazhong University of Science and Technology, China
Qun Jin	Waseda University, Japan
Arun Kumar Sangaiah	VIT University, India
Byung Seok Shin	Inha University, Korea

Publicity Chairs

Hyuk Jun Kwon	Soonchunhyang University, Korea
Jin Wang	Yangzhou University, China
Kwang-il Hwang	Incheon National University, Korea
Min Choi	Chungbuk National University, Korea
Hyuk Joon Kwon	Soonchunhyang University, Korea

Yeong-Seok Seo	Yeungnam University, Korea
Jinho Park	Dongguk University, Korea
Sushil Kumar Singh	SeoulTech, Korea
Abir El Azzaoui	SeoulTech, Korea

Program Committee

Ch. Z. Patrikakis	University of West Attica, Greece
Guillermo Camara Chavez	Universidade Federal de Minas Gerais, Brasil
Afrand Agah	West Chester University of Pennsylvania, USA
Francisco Jose Monaco	University of Sao Paulo, Brazil
Jin Kwak	Ajou University, Korea
Rainer Unland	University of Duisburg-Essen, Germany
Rajkumar Kannan	Affiliation Bishop Heber College, India
Fay Huang	National Ilan University, Taiwan
SungSuk Kim	Seokyeong University, Korea
Dongkyun Kim	KISTI, Korea
Yunhee Kang	Baekseok University, Korea
Chao-Tung Yang	Tunghai University, Taiwan
Ki-Sik Kong	Namseoul University, Korea
Xu Zhang	Chongqing University of Posts and Telecommunications, China
Waleed Farag	Indiana University of Pennsylvania, USA
Mingjie Liu	Chongqing University of Post and Telecom, China
Maumita Bhattacharya	Charles Sturt University, Australia
Huang Kuo-Chan	National Taichung University of Education, Taiwan
Wladyslaw Homenda	Warsaw University of Technology, Poland
Irene Chang	Shih-Hsin University, Taiwan
Yu-Chen Hu	Providence University, Taiwan
Bing Chen	Memorial University of Newfoundland, Canada
Ren-Song Ko	National Chung Cheng University, Taiwan
Joel Rodrigues	National Institute of Telecommunications, Brazil
Suren Byna	Lawrence Berkeley National Lab, USA
Abdel-Badeeh Salem	Ain Shams University, Egypt
Kyungbaek Kim	Chonnam National University, Republic of Korea
Jinli Cao	La Trobe University, Australia
Tatjana Davidovic	Serbian Academy of Sciences and Arts, Serbia
Caldelli Roberto	University of Florence, Italy
Koojoo Kwon	Baewha Women's University, Korea
Deok Gyu Lee	Seowon University, Korea

Jihoon Kan	Korea University, Korea
Hyun Woo Kim	Dongguk University, Korea
Jong Beom Lim	Pyeongtaek University, Korea
Yeong-Seok Seo	Yeungnam University, Korea
Ki Yong Lee	Sookmyung Women's University, Korea
Jong Hyuk Lee	Daegu Catholic University, Korea
Yeongwook Yang	Hanshin University, Korea

Contents

Classification of Diseases of Bean Plant Leaves Using Data Augmentation and Deep Neural Networks

Bui Hai Phong[1], Phuong Anh Nguyen[2], Tran Dang Quyet[1], Phung Duc An[3], Phan Duc Tri[4], and Le Anh Ngoc[2(✉)]

[1] Faculty of Information Technology, Hanoi Architectural University, Hanoi, Vietnam
phongbh@hau.edu.vn

[2] Swinburne Vietnam, FPT University, Hanoi, Vietnam
{anhnp75,ngocla2}@fe.edu.vn

[3] Wellspring Hanoi International Bilingual School, Hanoi, Vietnam

[4] Nguyen Sieu High School, Hanoi, Vietnam

Abstract. Bean plants have provided us with a rich source of food and industrial materials. However, diseases have significantly affected the production of bean plants. The recognition of bean plant diseases is an important step in the treatment of diseases. In recent years, deep neural networks have shown the leading performance for the image classification task. The paper presents a method for the recognition of diseased bean leaves. The vision of transformer (ViT) network has been applied to improve the accuracy of the recognition of diseased bean leaves. Moreover, data augmentation techniques have been applied to enlarge the image dataset to efficiently train the network. Obtained results of 97% for the classification of diseased bean leaves have shown the effectiveness of the proposed method.

Keywords: Bean plant leaf recognition · Disease classification · Machine learning · Deep learning · Data augmentation

1 Introduction

Bean plants have shown a significant value for our life and industry. However, diseases have significantly affected the productivity of bean plants every year [1]. Recognition of bean plant diseases allows us to diagnose and treat the diseases efficiently. Agricultural experts can recognize bean plant diseases based on the physical appearance of leaves. However, the method is time-consuming and requires much human effort. The automatic recognition of bean plant diseases aims to identify the diseases of beans using collected data without using experts. The automatic recognition of diseased bean leaves can be performed by machines to save time and human effort. However, the automatic recognition of bean diseases is a challenging task because there exist several difficulties of the recognition [2].

- The similarity of bean diseases may cause the errors of recognition.

J. S. Park et al. (Eds.): FutureTech 2025, LNEE 1483, pp. 1–8, 2025.
https://doi.org/10.1007/978-981-95-1999-6_1

- The challenges of condition of captured images of diseases may cause the errors of recognition.

Fig. 1. Examples of diseased bean leaves.

Figure 1 shows examples of diseased bean leaf images. Traditional recognition methods apply feature extraction techniques and machine learning algorithms [3] to recognize bean plant diseases. In recent years, deep neural networks (DNN) have shown the outstanding performance for the classification of images. The DNNs can extract features and classify images with high performance. The DNNs have applied to classify diseases of plant leaf images [3]. The paper presents a new method that applies the data augmentation technique and employs the Vision Transformation (ViT) network to improve the performance of the classification of bean leaf diseases [4]. The data augmentation plays an important role to increase and balance the number of bean leaf images that are necessary for the training and testing of DNN models. The ViT network has recently been proposed to significantly improve the accuracy of classification task. Furthermore, the paper compares the performance of various DNNs in the classification to analyze the strengths and weaknesses of the proposed method.

The rest of the paper is summarized as follows. A survey on research sources is presented in Sect. 2. Section 3 presents our proposed framework for the classification of diseases of bean. Experimental settings and obtained results are presented in Sect. 4. The last section of the paper gives the contributions and future works.

2 Related Works

This section provides the in-depth analysis of existing works for the classification of diseased bean leaves. Traditional methods extract visual features of plant leaf images, then various machine learning algorithms are applied to classify diseases [5]. In the early work of [6], the color and shape features are extracted to detect weeds in cereal fields. The Bayes rule and k-NN algorithms are proposed to detect weeds in images. The experiments are carried out on a small dataset with the accuracy of 79%. The work in [7] extracts color features of diseased citrus leaf images. Then, the statistic classification algorithms are applied to identify the citrus diseases. The method was tested on small dataset that consists of 40 citrus leaf images and 4 classes of diseases. The accuracy of the method is reported at 95%. The work in [8] extracted shape features of leaves to classify plant species. The work in [9] firstly extracts vein features of leaf images. Then, various machine learning algorithms are applied to classify legumes. In recent years, various deep neural network models are applied to improve the performance of the classification of plant diseases. The work in [2] fine-tuned CNN to classify diseased beans. The work in [10] applied U-net to detect bean leaves and then CNNs are applied to classify diseased beans. The work in [1] investigates various loss functions of CNNs to classify diseased bean leaves.

3 Proposed Method

The proposed system for the classification of bean leaves is described in Fig. 2. The system consists of the following modules: (1) First, the augmentation data using image processing is applied to increase and balance the number of images. (2) Second, the ViT network is applied to improve the classification accuracy. (3) Finally, the performance evaluation of the proposed method is carried out to highlight the effectiveness of the DNN model for the classification of diseased bean leaves.

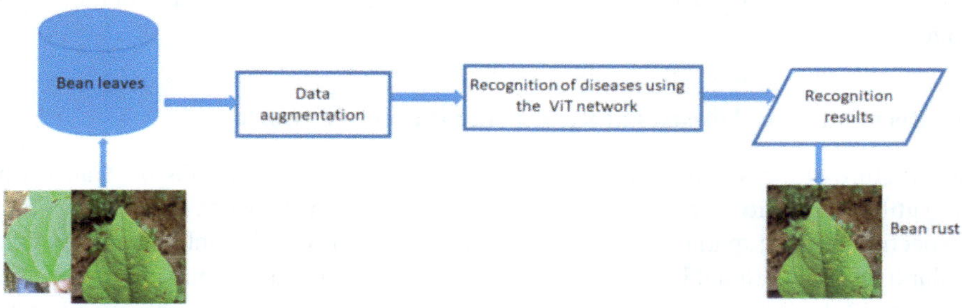

Fig. 2. The overall steps of our proposed method for the recognition of diseased bean leaves using the ViT network.

Fig. 3. Examples of the augmentation data of diseased bean leaves using image processing.

3.1 Data Augmentation

The number of bean images is a key factor to improve the performance of DNNs [2, 3]. To increase the number of bean images, the data augmentation using the image processing

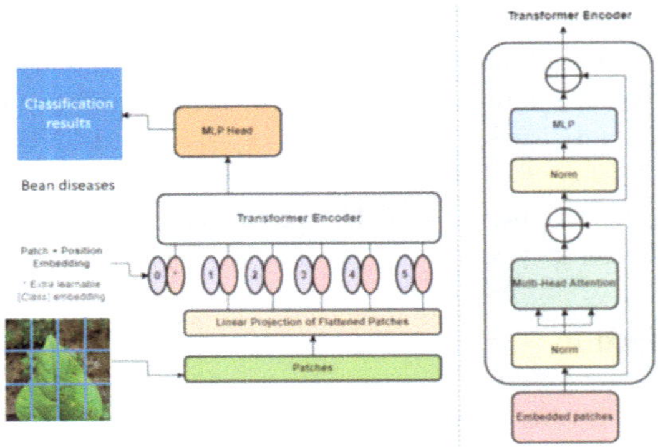

Fig. 4. Structure of ViT network for the classification of diseases of beans.

is applied. The rotation, transition and noise addition are applied to increase the number of input images. Figure 3 shows examples of original image and the data augmentation using image processing. The noise addition and the brightness adjustment of the original image.

3.2 Recognition of Diseases of Bean Using the ViT Network

The structure of the ViT network can be described as in Fig. 4. The Vision Transformer (ViT) utilizes attention mechanisms to establish connections between pixels in an image, irrespective of their spatial proximity. The input image is divided into smaller patches, similar to a convolutional layer with kernels. The patch embedding and position embedding convert the input images into flattened patches. Then, the patches are processed by the Transformer Encoder. The classification results are obtained using the MLP Head.

The ViT parameters in the training process are described as follows: The learning rate is set as 0.0001. The loss function is the cross-entropy.

The optimization solver is SGDM [11]. Figure 5 illustrates the loss and accuracy values during the training process of ViT network.

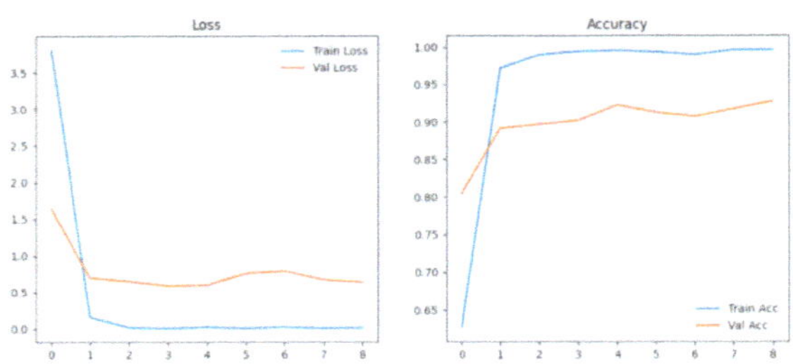

Fig. 5. Illustration of accuracy and loss values during training the ViT network of the classification diseases of beans.

4 Experimental Results

4.1 Dataset and Evaluation Metric

The proposed method has been evaluated on the public dataset of diseased bean leaves. The dataset consists of three types of diseases: Angular leaf spot, bean rust and healthy. We have applied data augmentation to improve and balance the number of bean leaf images. Detail information of the training and testing images of the datasets are shown in Tables 1 and 2, respectively. Eighty percent of bean leaf images is used for the training and the remainder is used for testing the DNNs. The accuracy metric that is commonly used in machine learning algorithms is applied to evaluate the performance of the proposed system [12]:

$$\text{Accuracy} = \frac{\text{Correct predictions}}{\text{Total predictions}} \tag{1}$$

Table 1. Detail information (number of images) of bean leaves in the dataset before applying the data augmentation.

Bean diseases	Training	Testing
Angular leaf spot	800	200
Bean rust	800	200
Healthy	800	200

Table 2. Detail information (number of images) of bean leaves in the dataset after applying the data augmentation.

Bean diseases	Training	Testing
Angular leaf spot	1600	400
Bean rust	1600	400
Healthy	1600	400

4.2 Performance Evaluation

The section analyzes and compares the accuracy of the proposed method with existed methods and other DNN models. Table 3 shows the performance of the DNNs for the classification of diseases of bean before applying the data augmentation. Table 4 shows that the classification of diseased bean leaves using the ViT network gains the highest accuracy compared to Resnet-50 [13] and Mobilenet V3 [14]. The data augmentation allows the DNNs obtain better results compared to that on the original data set. The data augmentation allows us obtain significant improvement of the classification accuracy. Table 5 shows that the execution time of ViT is better than Resnet-50. The Mobilenet V3 gains the highest score of execution time. Table 3 shows the accuracy of the classification of DNNs. ViT obtains highest accuracy compared to other DNNs. The proposed method obtained better results than the method in [1] because of the data augmentation techniques.

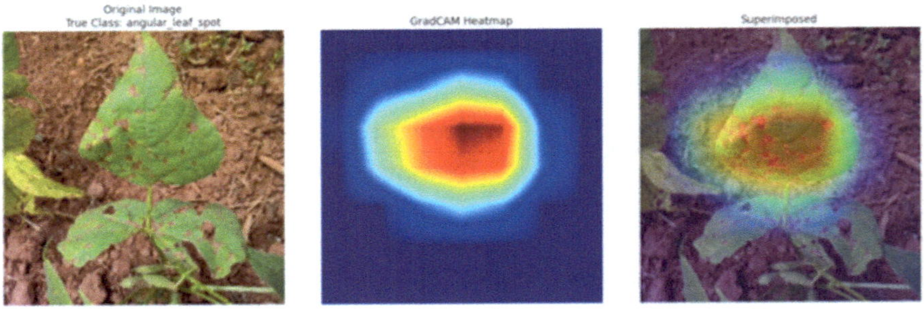

Fig. 6. Illustration of heat map obtained by the ViT network for the classification of diseased bean leaf images. The key features of original leaf images are remarked in red regions.

Fig. 7. Obtained results of the classification. The labels of images are classified correctly.

Figure 6 illustrates obtained heat map [15, 16] of the DNN. Figure shows that the DNN concentrates on the regions for the classification task. Figure 7 demonstrates the obtained results of our system. The figure demonstrates that the DNNs can classify the diseased leaf images in different conditions. The DNNs can classify diseases in tiny images or complex background efficiently.

Table 3. Performance comparison of the DNNs for the recognition of the diseased bean leaves before applying the data augmentation technique.

DNN model	Accuracy
Resnet-50	92%
Mobilenet V3	90%
ViT	**93%**

Table 4. Performance comparison of the DNNs for the recognition of the diseased bean leaves after applying the data augmentation technique

DNN model	Accuracy
Resnet-50	93%
Mobilenet V3	92%
EfficientNet	92%
ViT	97%

Table 5. Comparison of execution time of the DNNs for the recognition of the diseased bean leaves (in milliseconds)

DNN model	Execution time
Resnet-50	25
Mobilenet V3	**12**
ViT	35

5 Conclusion and Future Works

The paper presents the classification of diseased bean leaves using deep neural networks and data augmentation techniques. The ViT obtained the highest accuracy and the MobilenetV3 obtained the highest performance in term of execution time. The augmentation techniques are proposed to improve the accuracy of the classification. In the future, the augmentation using deep neural networks can be investigated to improve the accuracy. The results can be developed for smart agriculture applications.

References

1. Tavakoli, H., et al.: Leaf image-based classification of some common bean cultivars using discriminative convolutional neural networks. Comput. Electron. Agric. **181**, 105935 (2021). https://doi.org/10.1016/j.compag.2020.105935

2. Singh, V., et al.: Classification of beans leaf diseases using fine tuned CNN model. In: International Conference on Machine Learning and Data Engineering (2023). https://doi.org/10.1016/j.procs.2023.01.017
3. Phong, B.H.: Classification of plant leaf diseases using deep neural networks in color and grayscale images. J. Decis. Anal. Intell. Comput. **4** (2024). https://doi.org/10.31181/10002052024p.
4. Fan, R.-E., Chen, P.-H., Lin, C.-J.: Working set selection using second order information for training support vector machines. J. Mach. Learn. Res. **6**, 1889–1918 (2006)
5. Chen, X., Hsieh, C.-J., Gong, B.: When vision transformers outperform ResNets without pretraining or strong data augmentations. arXiv, https://arxiv.org/abs/2106.01548
6. Miller, W.M., Throop, J.A., Upchurch, B.L.: Pattern recognition models for spectral reflectance evaluation of apple blemishes. Post Harvest Biol. Technol. **14**, 11–20 (1998)
7. Perez, A.J., Lopez, F., Benlloch, J.V., Christensen, S.: Colour and shape analysis techniques for weed detection in cereal fields. Comput. Electron. Agric. **25**, 197–212 (2000)
8. Pydipati, R., Burks, T.F., Lee, W.S.: Identification of citrus disease using color texture features and discriminant analysis. Comput. Electron. Agr. **52**, 49–59 (2006)
9. Neto, J.C., Meyer, G.E., Jones, D.D., Samal, A.K.: Plant species identification using Elliptic Fourier leaf shape analysis. Comput. Electron. Agric. **50**(2), 121–134 (2006)
10. Larese, M.G., Namías, R., Craviotto, R.M., Arango, M.R., Gallo, C., Granitto, P.M.: Automatic classification of legumes using leaf vein image features. Pattern Recogn. **47**(1), 158–168 (2014). https://doi.org/10.1016/j.patcog.2013.06.012
11. Abed, S.H., Al-Waisy, A.S., Mohammed, H.J., et al.: A modern deep learning framework in robot vision for automated bean leaves diseases detection. Int. J. Intell. Robot. Appl. **5**, 235–251 (2021). https://doi.org/10.1007/s41315-021-00174-3
12. Mohammed, M.: Machine Learning: Algorithms and Applications. CRC Press (2016). https://doi.org/10.1201/9781315371658
13. He, K., Zhang, X., Ren, S., Sun, J.: Deep residual learning for image recognition (2016). https://arxiv.org/abs/1512.03385
14. Howard, A., et al.: Searching for MobileNetV3 (2019). https://arxiv.org/pdf/1905.02244v5
15. Selvaraju, R.R., Cogswell, M., Das, A., Vedantam, R., Parikh, D., Batra, D.: Grad-cam: visual explanations from deep networks via gradient-based localization. In: 2017 IEEE International Conference on Computer Vision (ICCV), pp. 618–626 (2017)
16. Basheer, I., Hajmeer, M.: Artificial neural networks: fundamentals, computing, design, and application. J. Microbiol. Methods **43**(1), 3 (2001)

Classification of Waste Bottles Using Image Processing and Deep Neural Networks

Hai Phong Bui[1] and Van Son Nguyen[2](✉)

[1] Faculty of Information Technology, Hanoi Architectural University, Hanoi, Vietnam
[2] Hanoi University of Science and Technology, Hanoi, Vietnam
sonnv@soict.hust.edu.vn

Abstract. Waste management is becoming an important issue to protect our environment. The issues in the field have attracted many researchers in recent years. The automatic classification of waste is one of the most important issues. In recent years, with the advances in computer vision and deep learning, the classification of waste has obtained promising results. The paper presents a method to classify waste bottles using image processing and advanced deep neural networks. The image processing aims to improve the quality of input images. The Vision of Transformer (ViT), an advanced deep neural network, is applied to improve classification accuracy. Moreover, dimensional reduction techniques are used to efficiently visualize the feature extraction of the deep neural networks. Obtained results of 91% for the classification of bottles and the performance comparison with existing methods has shown the effectiveness of our proposed method.

Keywords: Bottle Classification · Waste Management · Deep Neural Networks

1 Introduction

Waste management is becoming an important issue to protect our environment. The problems in the field have attracted many researchers in recent years [1]. The automatic classification of waste is one of the most critical issues. With the advances in computer vision and deep learning, waste classification has obtained promising results in recent years. The automatic classification of waste focuses on the discrimination of waste types using the analysis of collected data without using expert knowledge. An example of waste bottles is shown in Fig. 1. The paper investigates the automatic classification of wasted bottles, a well-researched topic [2]. However, the accurate classification of waste bottles is a challenging task.

The difficulties of the issue can be described as follows:

- The low quality of input images causes difficulties for the classification. The high variation of waste bottle shapes has caused classification difficulties.
- The similarity between the appearance of waste bottles causes difficulties in the classification.
- The deep learning models typically require large datasets. The lack of input images of waste bottles has caused difficulties in training deep-learning models efficiently.

J. S. Park et al. (Eds.): FutureTech 2025, LNEE 1483, pp. 9–15, 2025.
https://doi.org/10.1007/978-981-95-1999-6_2

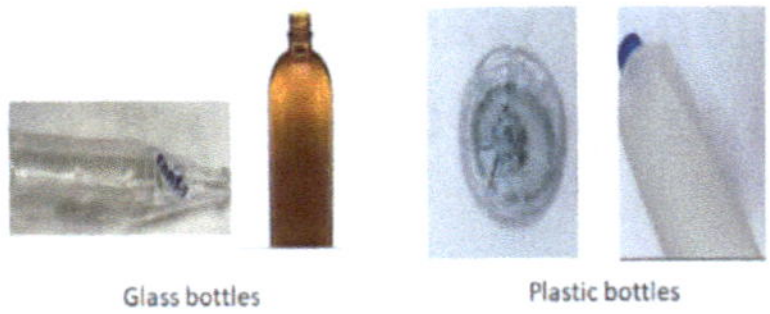

Fig. 1. Examples of various shapes and color bottles.

The paper presents a method to classify waste bottles using image processing and advanced deep neural networks. Input waste bottle images are pre-processed to improve the quality. After that, the ViT network [3] has been fine-tuned to improve the classification performance.

The rest of the paper is summarized as follows. A survey on research sources is presented in Sect. 2. Section 3 presents our proposed framework. Experimental settings and obtained results are presented in Sect. 4. The last section of the paper concludes the contributions and gives future works.

2 Related Works

This section reviews and analyzes existing approaches for the classification of waste. The work in [4] applied the ANN model to analyze the features of waste bottles. The work in [5] develops ANN-based models to predict municipal waste generation. The work in [6] applied the Resnet-50 and Support Vector Machine (SVM) to classify waste images. In the recent research in [7], the YOLOv10 and Mobilenetv2 were applied to classify plastic waste. The method is evaluated on the dataset that consists of 4000 plastic waste images. The work in [8] applied the Mobilenetv2 network to classify waste materials. As the analysis, the classification of waste using deep learning methods have achieved promising results, however, the issue needs to be further investigated to improve the performance and real waste datasets.

3 Proposed Method

The main steps of our proposed method are shown in Fig. 2. The framework consists of the following steps: image processing, data augmentation and classification using DNNs. First, input images are normalized at 384×384. Then, the data augmentation is applied using the image processing techniques: rotation, noise addition, and reflection. The data augmentation technique allows to improve the performance of the DNNs. Figure 3 shows examples of the data augmentation using image processing techniques. The data augmentation allows the enhancement and robustness of DNNs [9,12]. The translation transformation, rotation, and noise addition are applied to increase the number of original bottle images.

The structure of ViT is shown in Fig. 4. The network consists of two main components: the backbone and the head. The backbone is responsible for the encoding step of the network. The backbone takes the input images and outputs a vector of features. The head is responsible for making the predictions. The head maps the encoded feature

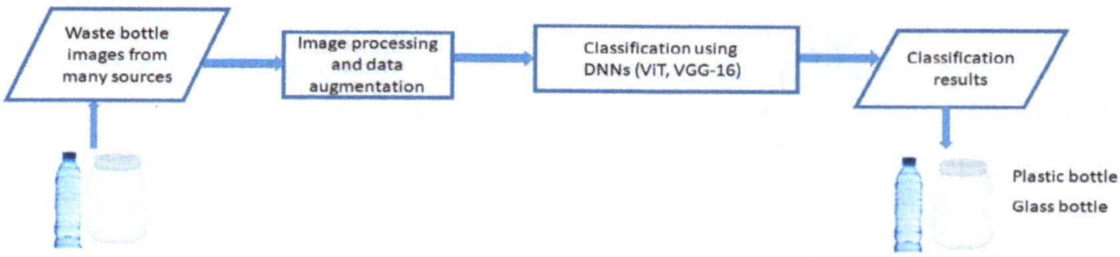

Fig. 2. The overall steps of our proposed method for the classification of waste bottles.

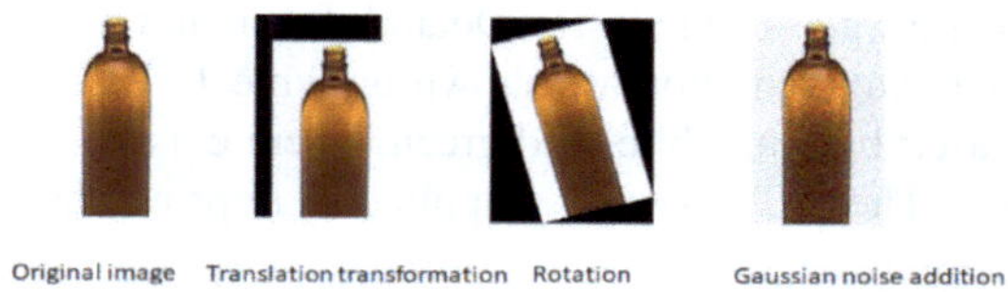

Fig. 3. The structure of the ViT network for the classification of waste bottles.

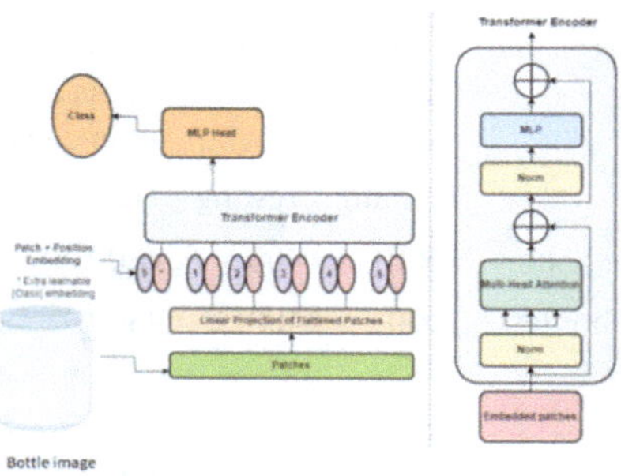

Fig. 4. Structure of ViT network for the classification of bottles

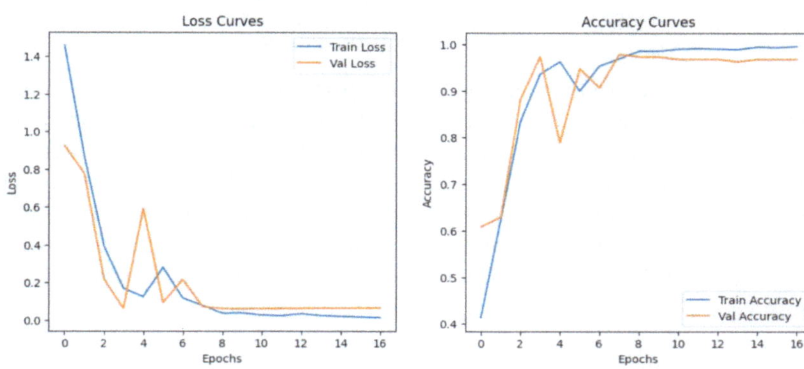

Fig. 5. Illustration of accuracy and loss values during training the ViT network of the classification of bottle images.

vectors to the prediction scores. The network processes the patch size of 16 of input images [10]. The AdamW optimizer algorithm is applied for the training of the network. The learning rate is set as 0.0001 and the max epoch is 20. The accuracy and loss values of the training process are shown in Fig. 5.

4 Performance Evaluation

The proposed method has been evaluated on the public data set of waste bottles. There exist two types of bottles: glass and plastic. Detail information of the dataset before and after applying the data augmentation are shown in Table 1. There exist various colors of bottles in the dataset: brown, white, and green. There exist one or several bottles in images in the dataset. The ViT network is applied to improve the performance of the classification.

Table 1. Number of bottle images used for the training and testing DNNs.

Dataset	Before applying the data augmentation		After applying the data augmentation	
	Training	Testing	Training	Testing
Plastic bottles	1600	400	4800	1200
Glass bottles	1600	400	4800	1200

We apply the precision (P), recall (R) and F1 score metrics to evaluate the performance of the proposed method [11]:

$$\mathrm{F1-score} = \frac{2 \times \mathrm{Precision} \times \mathrm{Recall}}{\mathrm{Precision} + \mathrm{Recal}}$$

In this section, we evaluate and analyze the performance of the proposed method. As shown in Tables 3 and 4, the ViT network obtained the highest classification accuracy compared to other networks. The Resnet-512 [13] extracts features better than the VGG-16 [14] because the network consists of more layers. The classification accuracy has slightly improved after applying the data augmentation. Actually, the ViT network applies self-attention to sequences of image patches instead of using convolution layers as in CNNs. Compared to the classification method in [6], the application of DNNs gains better results.

The application of PCA and SVM obtains lower accuracy. However, the computational time of the method is better than other methods. The PCA technique allows us to reduce the number of features extracted by CNN. Therefore, the computational time is better compared to other methods. Table 2 shows the comparison of the execution time of the classification methods of bottle images.

Figure 6 illustrates the feature distribution using the PCA technique. In this case, the Resnet-512 extracted 512 features of images. The PCA reduces the number of features to two. The PCA allows the efficient visualization of the extracted feature in 2D dimensions. Figure 7 shows the confusion matrix that represents the performance of our classification model with the feature extraction. The confusion matrix shows that we can classify bottles with high accuracy. The misclassification may come from the following reasons: low quality of images caused the errors and/or various overlapped objects in images caused the errors (Table 2).

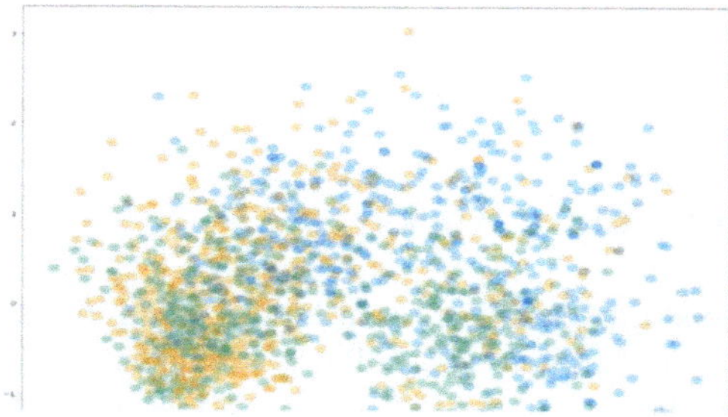

Fig. 6. The visualization of features extracted by DNNs using the PCA technique. Extracted features of plastic and glass bottles are denoted by yellow and blue colors, respectively.

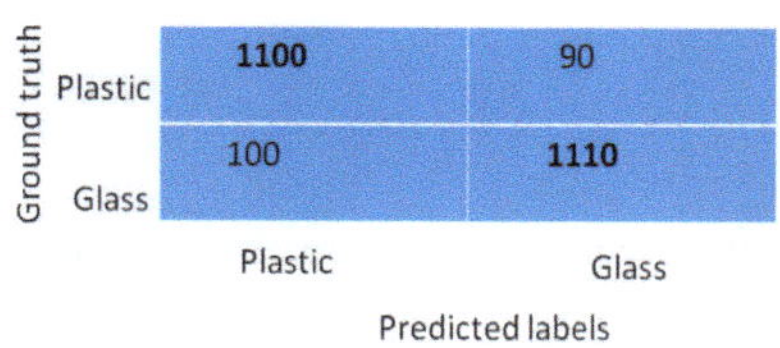

Fig. 7. Confusion matrix of our classification model using DNNs.

Table 2. Performance comparison before applying the image processing and data augmentation

Methods	P	R	F1
VGG-16 [14]	84%	83%	83.5%
Resnet-50 [13]	86%	84%	85%
ViT	90%	88%	88.98%
PCA and SVM [6]	80%	78%	79%

Table 3. Performance comparison after applying the image processing and data augmentation

Methods	P	R	F1
VGG-16 [14]	86%	85%	85.5%
Resnet-50 [13]	89%	87%	87.98%
ViT	91%	90%	90.5%
PCA and SVM [6]	83%	81%	82%

Table 4. Comparison of execution time of the classification methods on waste bottles

Methods	VGG-16	Resnet-50	ViT	PCA & SVM
Execution time (milliseconds)	13	26	54	8

5 Conclusion and Future Works

The paper has presented the classification of bottles using image processing and deep neural networks. The image processing allows to improve the quality of training an testing images. The application of the ViT network allows us to obtain higher accuracy. The PCA technique is applied to reduce the number of extracted features of bottles and we can visualize the features efficiently. In the future, the results can be developed to manage waste efficiently to protect our environment.

References

1. Fang, B., et al.: Artificial intelligence for waste management in smart cities: a review. Environ. Chem. Lett. **21**(4), 1959–1989 (2023)
2. Liu, K., Liu, X.: Recycling material classification using convolutional neural networks. In: 2022 21st IEEE International Conference on Machine Learning and Applications (ICMLA), pp. 83–88 (2022)
3. Chen, X., Hsieh, C.-J., Gong, B.: When vision transformers outperform ResNets without pretraining or strong data augmentations. arXiv (2022)
4. Batinic, B., Vukmirovic, S., Vujić, G., Stanisavljević, N., Ubavin, D., Vukmirović, G.: Using ANN model to determine future waste characteristics in order to achieve specific waste management targets case study of Serbia. J. Sci. Ind. Res. (2011)
5. Antanasijević, D., et al.: The forecasting of municipal waste generation using artificial neural networks and sustainability indicators. Sustain. Sci. **8**, 37–46 (2013)
6. Adedeji, O., Wang, Z.: Intelligent waste classification system using deep learning convolutional neural network. In: 2nd International Conference on Sustainable Materials Processing and Manufacturing, vol. 35, pp. 607–612 (2019)
7. Kunwar, S., Owabumoye, B.R., Alade, A.S.: Plastic waste classification using deep learning: insights from the WaDaBa dataset. arXiv preprint arXiv:2412.20232 (2024)

8. Selim, S., Ecemis, I.N.: Classification of urban waste materials with deep learning architectures. SN Comput. Sci. (2023)
9. Gracia Moisés, A., Vitoria Pascual, I., Imas González, J.J., Ruiz Zamarreño, C.: Data augmentation techniques for machine learning applied to optical spectroscopy datasets in agrifood applications: a comprehensive review. Sensors **23**(20) (2023)
10. Dosovitskiy, A.: An image is worth 16x16 words: transformers for image recognition at scale. arXiv preprint arXiv:2010.11929 (2020)
11. Barbedo, J.G.A.: A review on the main challenges in automatic plant disease identification based on visible range images. Biosyst. Eng. **144**, 52–60 (2016)
12. Ren, S., He, K., Girshick, R., Sun, J.: Faster R-CNN: towards real-time object detection with region proposal networks. In: NIPS, pp. 91–99 (2015)
13. He, K., et al.: Deep residual learning for image recognition. In: Proceedings of the IEEE Conference on Computer Vision and Pattern Recognition (2016)
14. Simonyan, K.: Very deep convolutional networks for large-scale image recognition. arXiv preprint arXiv:1409.1556 (2014)

Integrating NLP Platform to Build a Chatbot in the Hospitech System

Tam Dinh Thi[2], Thang Ta Quoc[1], Khoai Vu Tien[1], Huyen Trinh Thi Thu[1], Thuy Nguyen Thu[1], Thai Nguyen Dinh[1], Hanh Nguyen Thi Phuong[1], and Trung Tran[1](✉)

[1] Electric Power University, Hanoi, Vietnam
{thaind,hanhntp,trungt}@epu.edu.vn
[2] Ho Chi Minh City University of Industry and Trade, Ho Chi Minh City, Vietnam
tamdt@huit.edu.vn

Abstract. Integrating an NLP platform into the Hospitech system leverages natural language processing (NLP) technology to develop chatbotAI capable of analyzing symptoms and providing online health consultations. The chatbot comprehends context and delivers accurate and flexible responses, improving communication quality between patients and doctors. It enables users to access information, schedule appointments, and receive support efficiently. With the ability to learn from new data, the chatbot continuously enhances accuracy and better meets patient needs. This reduces the workload of medical staff, optimizes schedule management, and improves user experience.

Keywords: Hospitech · AI Chatbot · Natural Language Processing (NLP) · Online Health Consultation · Online Appointment Scheduling · Healthcare Process Optimization

1 Introduction

Amid the digital transformation of the healthcare sector to meet increasing care demands, the project "Integrating an NLP Platform to Build a Chatbot in the Hospitech System" aims to deliver an advanced solution for analyzing user symptoms and providing quick, accurate diagnostic suggestions. This is not merely an online appointment scheduling tool but also a virtual healthcare assistant, optimizing interactions between patients and doctors.

Built on natural language processing (NLP) and artificial intelligence (AI), the Hospitech chatbot is designed to understand and process users' complex symptom descriptions. With context analysis capabilities, the chatbot not only suggests potential medical conditions but also provides useful information to assist doctors in making accurate diagnoses. Beyond scheduling appointments, the chatbot enables patients to access healthcare services comprehensively, quickly, and conveniently.

The system enhances interaction and communication between patients and doctors. The chatbot can answer common questions, offer basic medical advice, and provide

J. S. Park et al. (Eds.): FutureTech 2025, LNEE 1483, pp. 16–23, 2025.
https://doi.org/10.1007/978-981-95-1999-6_3

detailed information on the examination process. This not only reassures patients and supports them but also reduces waiting times and improves doctors' work efficiency.

The project aims to establish a modern, seamless online healthcare platform that caters to everyone's healthcare needs anytime, anywhere. By integrating NLP and AI, Hospitech not only supports swift diagnosis but also lays the foundation for a smart healthcare ecosystem where technology collaborates with humans to provide the best care possible.

2 Input Data and Processing Workflow

2.1 Data Sources

- Hospital Data – Information about departments, doctors' schedules, and official medical data.
- User Data – Questions, personal registration details, and consultation history.
- Chatbot Data – Interaction history and performance evaluations.
- Medical Resources – Scientific articles and official guidelines.

2.2 Types of Collected Data

- Text Data: Medical information in text formats (Text, Word, Excel).
- Image Data: Illustrative pathology images.
- URL Data: Links to reliable sources.

2.3 Data Processing Workflow

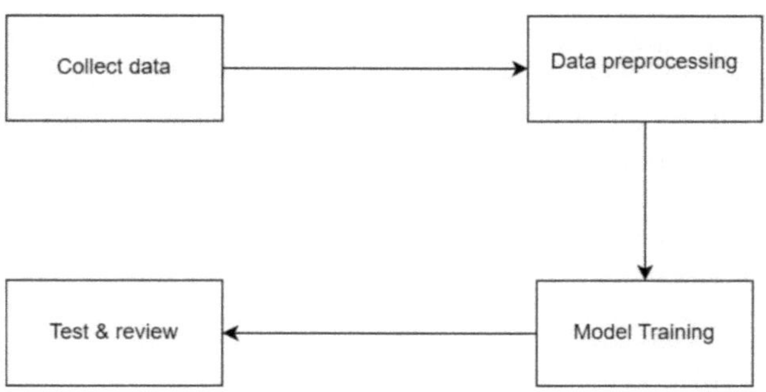

Fig. 1. Data Processing Workflow Diagram

Explanation: Data Collection: From hospitals, users, the chatbot, and reliable medical sources; Data Preprocessing: Removing noise and normalizing data; Model Training: Using the Transformer model; Testing & Evaluation: Assessing accuracy, communication ability, and performance (Fig. 1).

Data collection gathers information from various sources, including hospitals, users, the chatbot, and trusted medical sources. This information may include medical data, user queries, consultation history, and scientific articles. Collection is performed via APIs and online protocols, ensuring automated and continuous information intake.

Subsequently, data preprocessing ensures information accuracy and quality. This involves noise removal, such as redundant or irrelevant data, and normalization, such as converting inconsistent formats into a unified standard and eliminating errors in text. This process creates clean datasets for further analysis and machine learning model processing.

2.4 Input Data Structure Specification

The AI chatbot's input data structure is specified to ensure accuracy and consistency in receiving and processing information. The system supports three main data types: text, URLs, and files.

Text data is used to train the chatbot on symptoms, diseases, and medical terminology. This enables the chatbot to accurately understand and analyze user requests. These text samples help the system learn to identify sentence patterns and keywords related to symptoms or conditions, thereby enhancing the chatbot's ability to classify and process medical information effectively.

URLs are utilized to link reputable medical resources, such as disease databases or professional articles. These URLs must ensure authenticity and reliability to maintain the system's accuracy.

Files support popular formats like Excel, Word, and PDF. These files store detailed information about diseases, symptoms, or basic profiles of doctors. For example, an Excel file must adhere to a fixed structure, with data columns such as disease name, primary symptoms, and prevalence rates. Word or PDF files often contain medical guidelines or research reports, providing the chatbot with comprehensive and diverse reference materials. (Include suggestions for doctor appointment scheduling).

Input data is submitted via three user interface methods: direct text entry, file uploads, or URL provision. The system validates the data, rejecting invalid entries (e.g., incorrect file formats or broken URLs) and requesting users to resubmit. Restrictions, such as file size limits under 10MB, are applied to ensure operational efficiency.

The detailed specification of input data structure plays a crucial role in optimizing the chatbot's effectiveness, ensuring it provides precise information and healthcare advice tailored to user needs.

3 AI Model and Training Workflow

3.1 Machine Learning Model Description

The machine learning model applied in this system is the Transformer, a state-of-the-art architecture in natural language processing (NLP). Its attention mechanism processes and analyzes relationships between words in a sentence, unaffected by word order, enabling it to deeply understand sentence semantics even in complex structures. This capability ensures outstanding performance in tasks like language translation, text summarization, and question-answering.

By leveraging the Transformer's superior features, the chatbot enhances communication quality between patients and the system, optimizing consultation processes and delivering convenience and efficiency for users.

3.2 Training Process

The data processing and model training process includes several steps: data preparation, model initialization, training, evaluation, and fine-tuning.

After data collection, the next step is data preprocessing, a critical process to eliminate noise and standardize the data, ensuring the quality of the input information. This step helps prevent the model from being skewed or influenced by inaccurate data while improving the efficiency of the training process.

Once the data has been processed, a Transformer model will be trained using the prepared dataset. The training process is conducted with optimization techniques to enable the model to understand and respond accurately to user context and requests. High-quality training data ensures the model achieves effective consultation and question-answering capabilities.

Finally, after the training process is complete, the model will undergo testing and evaluation. These tests assess the accuracy of its responses and its ability to understand the semantics of the text. This process helps determine the model's success level and identify areas that require adjustment to enhance performance.

3.3 The Process of Creating ChatbotAI

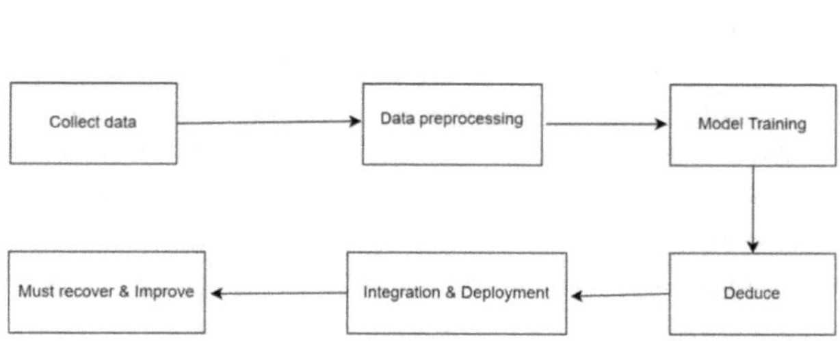

Fig. 2. Diagram of the AI Chatbot Creation Process

Consists of six main steps: (i) Data Collection: Collect diverse data from sources such as medical documents, conversational scripts, and expert evaluations, storing it in a properly structured database; (ii) Data Preparation: Clean the data, standardize language and formats, then transform the data into numerical vectors for easier model processing; (iii) Model Training: Utilize architectures like GPT, BERT, or BioBERT to train the model on preprocessed datasets, optimizing to improve performance; (iv) Inference System: The model analyzes user requests, connects to the database, and generates accurate context-based responses; (v) Deployment: Combine the model with user interfaces (UI) and APIs, deploying it across platforms such as websites, mobile apps, or messaging services; (vi) Feedback and Improvement: Collect real-world feedback, analyze it, and retrain the model to enhance the chatbot's accuracy and efficiency (Fig. 2).

3.4 Chatbot Training Function

The Hospitech ChatbotAI is trained through three primary methods: text, URLs, and files (Excel, Word, PDF). Each method plays a crucial role in providing in-depth knowledge and supporting users effectively.

Text data is used to train the chatbot with information on symptoms, diseases, and medical terminology. This allows the chatbot to accurately understand and analyze user requests.

Data sourced from URLs provides continuous information from reputable websites and the latest updates in healthcare. This ensures that the system maintains high accuracy and reliability in its responses.

Additionally, files (Excel, Word, PDF) store detailed information about diseases, symptoms, and doctor profiles. This enables the chatbot to offer more personalized and specific advice tailored to the unique characteristics of each patient and the expertise of the doctors.

All these data sources combined help the chatbot not only deliver automated responses but also create a professional and user-friendly medical consultation experience. This integration also enhances the accuracy of diagnoses and improves user support.

4 Testing and Evaluation

4.1 Testing

The testing process focuses on checking and evaluating the features of the system. First, the accuracy of the chatbot's responses will be tested by comparing them with verified medical information. Next, the chatbot's natural communication ability will be assessed through experimental conversations to evaluate its understanding and appropriate responses in various scenarios. Finally, the system will be tested for performance, including its ability to handle multiple resources and ensure stable response speed when many users interact simultaneously (Tables 1 and 2).

4.2 Effectiveness Evaluation

The effectiveness of the system is evaluated through user feedback. Surveys show that 90% of users rate the chatbot highly for providing accurate medical information and its natural communication ability. This demonstrates that the system meets user expectations by delivering medical information quickly, clearly, and understandably, while also improving the user experience when using the service (Fig. 3).

Table 1. Accuracy.

No	Q	A	Chatbot'A	R/W	Presicion	Recall	F1-Score
1	"Bệnh tim là gì?"	"Bệnh tim là tình trạng mà tim không thể hoạt động đúng cách, có thể dẫn đến đau ngực, khó thở,…"	"Bệnh tim là tình trạng ảnh hưởng đến hoạt động của tim"	R	1	1	1
2	"Nguyên nhân gây ra bệnh tim?"	"Các nguyên nhân chính bao gồm huyết áp cao, cholesterol cao, tiểu đường, …"	"Huyết áp cao và cholesterol là nguyên nhân chính gây bệnh tim"	T	1	0,9	0,95
3	"Bệnh tim có chữa được không?"	"Bệnh tim có thể kiểm soát thông qua thuốc và thay đổi lối sống"	"Bệnh tim có thể kiểm soát thông qua thuốc và chế độ ăn uống"	R	1	0,9	0,95
4	"Bệnh tim có di truyền không?"	"Có bệnh tim có thể di truyền từ cha mẹ"	"Bệnh tim có thể di truyền từ cha mẹ"	R	1	1	1
5	"Những triệu chứng của bệnh tim là gì?"	"Đau ngực, khó thở, mệt mỏi, nhịp tim không đều"	'Triệu chứng bệnh tim bao gồm đau ngực và khó thở"	R	1	0,8	0,89

Table 2. Training Symptoms.

No	Disease name	Symptom	Treatment	Risks and complications	Severity
1	Cảm cúm	Sốt, ho, sổ mũi. đau cơ	Nghỉ ngơi, uống thuốc giảm đau, uống nước nhiều	Biến chứng viêm phổi, suy hô hấp	Nhẹ
2	Bệnh tim	Đau ngực, khó thở, mệt mỏi, chóng mặt	Thuốc, thay đổi chế độ ăn uống, phẫu thuật	Đột quỵ, suy tim. nhồi máu cơ tim	Nghiêm trọng

(*continued*)

Table 2. (*continued*)

No	Disease name	Symptom	Treatment	Risks and complications	Severity
3	Tiểu đường	Khát mước, đi tiểu nhiều, mệt mỏi	Kiểm soát đường huyết, thuốc, chế độ ăn uống	Biến chứng thận, mù lòa, tai biến mạch máu	Nhẹ đến vừa phải
4	Viêm phổi	Ho, sốt, khó thở, đau ngực	Kháng sinh, nghỉ ngơi, uống nhiều nước	Suy hô hấp, nhiễm trùng máu	Trung bình đến nặng
5	Cảm lạnh	Nghẹt mũi, hắt hơi, đau họng	Nghỉ ngơi, thuốc giảm đau, uống nước ấm	Biến chứng viêm tai, viêm phế quản	Nhẹ

Fig. 3. System Testing Results

5 Conclusion and Future Prospects

The integration of the NLP platform to build the chatbot in the Hospitech system has yielded remarkable results. The chatbot has reduced the volume of calls to the hotline by 30%, saving patients time and reducing stress when scheduling appointments.

In the future, we will continue to develop the chatbot to further personalize the user experience by using machine learning algorithms to analyze interaction history and provide relevant recommendations. Additionally, integrating the chatbot with wearable medical devices will allow for continuous health monitoring and early detection of abnormal signs. We are committed to ensuring the utmost security of patients' personal information and continuously upgrading the chatbot to meet the growing needs of users. Our ultimate goal is to build a chatbot that becomes a reliable companion, assisting patients in better managing their health and improving their quality of life.

References

1. An, P.V.: Mạng lưới cảm biến không dây trong theo dõi bệnh nhân. Tạp chí Công nghệ Sức khỏe (NXB Công nghệ), Hà Nội, 30–38 (2019)
2. Ayanouz, S., Abdelhakim, B.A., Benhmed, M.: A smart chatbot architecture based on NLP and machine learning for health care assistance. In: Proceedings of the 3rd International Conference on Networking Information Systems & Security (2020)

3. Athota, L., et al.: Chatbot for healthcare system using artificial intelligence. In: 2020 8th International Conference on Reliability, Infocom Technologies and Optimization (Trends and Future Directions) (ICRITO), Noida, India, pp. 619–622 (2020)
4. Bình, N.V.: Ứng dụng di động trong quản lý sức khỏe cá nhân. Tạp chí Tin học Y tế (NXB Tin học), TP. Hồ Chí Minh, 40–48 (2021)
5. Kumar, M.S., et al.: Medbot-medical diagnosis system using artificial intelligence. EAI Endorsed Trans. Smart Cities **6**(17), 1–5 (2021)
6. Long, T.V.: Phát triển hệ thống quản lý bệnh viện bằng công nghệ blockchain. Tạp chí Khoa học và Công nghệ (NXB Khoa học), TP. Hồ Chí Minh, 10–17 (2022)
7. Mai, L. T.: Sử dụng dữ liệu lớn trong phân tích y học. Tạp chí Y tế (NXB Y học), Đà Nẵng, 25–32 (2020)
8. Patil, A.: Healthcare chatbot using artificial intelligence. J. Res. Appl. Sci. Eng. Technol. **10**(8), 1–7 (2022)
9. Xu, L., et al.: Chatbot for health care and oncology applications using artificial intelligence and machine learning: systematic review. JMIR Cancer **7**(4), 1–18 (2021)
10. Uyên, N.T., Văn, P.T., Thanh, H.T.: ỨNG DỤNG TRÍ TUỆ NHÂN TẠO TRONG ĐÁNH GIÁ BỆNH VÕNG MẠC ĐÁI THÁO ĐƯỜNG TẠI BỆNH VIỆN ĐA KHOA HÀ ĐÔNG. Tạp Chí Y học Việt Nam 536(2) (2024)

A Comparative Study of LGPMA and Table Transformer in Table Structure Recognition

Tang Van Nguyen[1(✉)], Dang Hai Bui[1], and Phuong Anh Nguyen[2]

[1] Faculty of Data Science and Technology, Foreign Trade University, Hanoi, Vietnam
tangnv@ftu.edu.vn

[2] Swinburne University of Technology, Hanoi, Vietnam

Abstract. Currently, there are numerous approaches to tackle the problem of recognizing table structures in unstructured documents, in which deep learning stands out as an efficient one. However, applying the deep learning algorithms in real-world scenarios poses significant challenges due to the diversity of tables found in documents evolving daily.

Therefore, researching and comparing various TSR (TSR) models for identifying tables in practical documents is not only academically significant but also holds practical implications for applying Artificial Intelligence (AI) to real-world scenarios. Discovering a suitable or superior model for table recognition can help research groups and businesses save time, costs, and optimize product development pathways. In this paper, we perform a comparative study between the two state-of-the-art models through the years, namely Local and Global Pyramid Mask Alignment (LGPMA) and Table Transformer (TATR), using two datasets that we created to represent practical tables. These datasets include the Borderless Table dataset, representing tables commonly found in research articles, and the Vietnamese Tax Liability dataset. To evaluate the performance of the above models, we used the Tree-Edit-Distance-Based Similarity metric, considering only the table structure (TEDS-Struc.). Finally, we draw conclusions and suggest some recommendations for the practical applications of these methods.

Keywords: Table Structure Recognition (TSR) · Table Transformer (TATR) · Local and Global Pyramid Mask (LGPMA) · Tree Edit Distance Based Similarity (TEDS)

1 Introduction

The problems of extracting and understanding meaningful information and insights from digital documents have received attention in the past decades. A digital document typically consists of several main components, including figures, text, and tables. Tables play an important role in presenting structured information in a clear and comprehensible representation. Then, Table Structure Recognition (TSR), a critical task in document analysis applications, can facilitate table recognition and extraction for further process.

A table is a structured representation of data organized into rows and columns. Moreover, a table structure with spanning cells, partially bordered or even none-bordered, may

J. S. Park et al. (Eds.): FutureTech 2025, LNEE 1483, pp. 24–30, 2025.
https://doi.org/10.1007/978-981-95-1999-6_4

introduce additional complexities in TSR. There are several common approaches dealing with TSR. Deep learning [1] has been proven to be very effective in table detection and TSR. However, there is no method that completely solves all types of tables in real-world unstructured documents. Therefore, researching and comparing various TSR models for identifying tables in practical documents is not only academically significant but also holds practical implications for applying Artificial Intelligence (AI) to real-world scenarios. In this paper, we perform a comparative study between the two state-of-the art models through the years, namely Local and Global Pyramid Mask Alignment (LGPMA) [2] and Table Transformer (TATR) [3], using two datasets that we created to represent practical tables. These datasets include the Borderless Table Dataset, representing tables commonly found in research articles, and the Vietnamese Tax Liability Dataset, representing scanned tables from a specific country. The evaluation metric used is Tree-Edit-Distance-based Similarity (TEDS). The experiments showed that TATR outperformed LGPMA on unseen data, especially the Vietnamese Tax Liability dataset, but was less accurate on the Borderless Table dataset. The remainder of this paper is as follows: Sect. 1 shows a brief introduction to TSR. Section 2 describes the related work on LGPMA and TATR. Section 3 is devoted to the data analysis and experiments. Finally, Sect. 4 shows the conclusion and perspectives.

2 Related Work

TSR models fall under several categories. The most common of these are object detection, semantic image segmentation, recurrent neural networks, deformable or dilated convolutions, and graph neural networks. In 1998, Kieninger and Dengel [4] introduced the preliminary methodology for TSR, wherein they grouped the text into chunks and further subdivided these chunks into cells based on the column borders. Recent deep learning-based approaches, such as Sebastian Schreiber, et al. [1], employed Faster RCNN for table detection and, subsequently, applied deep learning-based semantic segmentation for TSR.

The models we tested, LGPMA and TATR, can be classified as image segmentation and object detection, respectively. The latter of these has been derived from the DETR model [5], which utilizes the transformer architecture to obtain bounding boxes and class labels from images. The former is built upon Mask R-CNN, which uses a region-based convolutional neural network to produce segmentation masks along with their corresponding labels. Grid boundaries are detected via the segmentation process, which comprises the global method. The local method consists of a bounding box detection task for the individual cells.

The evaluation metric from [6] calculates adjacency relations between non-empty cells but has limitations in detecting errors involving empty cells and assessing the finer details of cell content recognition. To overcome these limitations, we used evaluation metric first introduced in [7], called Tree-Edit Distance-based Similarity (TEDS), which evaluates table structures globally and measures cell content recognition through string-edit distance.

However, considering OCR errors in the evaluation can lead to unfair comparisons due to the variability in performance across different OCR models. To address this, recent work [2] has introduced a modified version of the TEDS metric, known as TEDS-Struc., which focuses solely on assessing the accuracy of TSR while ignoring OCR errors. For our evaluation on our datasets, we utilize this modified TEDS-Struc. metric to ensure a fair assessment of structural recognition performance.

3 Implementation and Experiment

3.1 Datasets Description

Recently, many table datasets exist for table recognition, but this study focuses on those with markup annotations (HTML, XML) [3, 5]. However, inconsistencies and annotation errors [3] can affect evaluation and underestimate model performance. To address this, two small, carefully curated and consistently labeled datasets were created.

Borderless Tables Dataset. The first dataset was intended to evaluate the performance of the mentioned models when predicting special structure tables, especially borderless tables. This dataset has 100 tables which are selected from PubTables-1M [3] with some conditions; for instance, the structure of all the tables had merged cells, the borders surrounding every cell were invisible as much as possible. The reasons for choosing the PubTables-1M dataset are that it fixed missing annotation information to the maximum and made the annotations consistent [3].

Vietnamese Tax Liability Dataset. The second dataset included the tables that were collected from Vietnamese Tax Liability documents. The purpose of this dataset is to evaluate the accuracy of these models on unseen data.

The documents consisted of 272 papers that were scanned into PDF files. The processing included pre-processing image techniques and data analysis, which were applied to these PDF files to obtain the images of Vietnamese Tax Liability tables. Blank and non-table pages were removed, misoriented pages were corrected, and surrounding spaces and text were cropped to minimize their impact on TSR model inferences. We give an example of before and after the processing of one image in this dataset (See Fig. 1).

As a result, the Vietnamese Tax Liability dataset contains 258 tables. The tables in this dataset have the following characteristics: they often contain multi-row headers, include merged cells and merged rows, and some tables have a large number of closely spaced rows. The ground truth annotations manually created in HTML format, following the PubTabNet [7] dataset format. Figure 2 shows an HTML excerpt for a table with a merged cell.

3.2 Implementation Detail

All experiments are implemented in Pytorch with NVIDIA Tesla A30-24 GB GPU. Two pre-trained models of TATR and LGPMA were used. Due to the incomplete labeling of the Vietnamese tax dataset, we initially utilize pre trained models for testing and evaluation. The pre-trained TATR model on the combination of PubTables-1M and FinTabNet [8] datasets applied the Detection Transformer (DETR) and used a ResNet-18 backbone pre-trained on ImageNet with the first few layers frozen. The pre-trained LGPMA model

(a) Original PDF page (b) Table image

Fig. 1. The example of before and after the processing one table in the Vietnamese Tax Liability Table Dataset

on the PubTabNet dataset used a backbone of ResNet-50 with Feature Pyramid Network (FPN). The weights of the backbone were initialized from the pre-trained model of MS COCO [2].

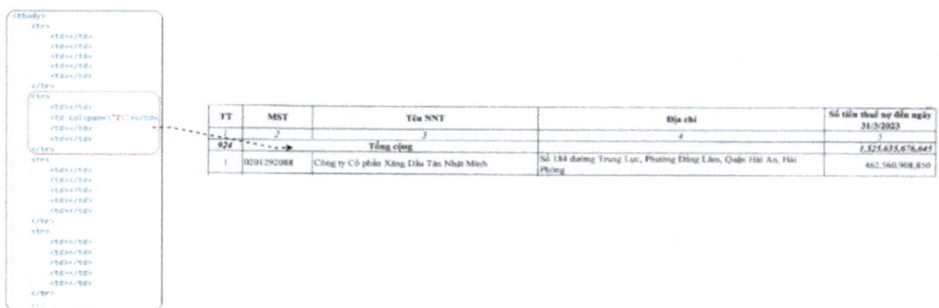

Fig. 2. The example of an HTML file corresponding to the table image

3.3 Results and Comparison

The results of testing TATR and LGPMA models on two datasets that were introduced in Subsect. 3.1. We assessed the effectiveness of two models in terms of TEDS-Struc. metric scores. The scores on the TEDS-Struc. metric are shown in Table 1. The TATR model predicts the results less accurately than the LGPMA model on the Borderless Table dataset, specifically scoring lower by 7.39%. However, on the Vietnamese Tax Liability dataset, the TATR model demonstrates superiority over the LGPMA model, with a score difference of 36.15%.

To conduct a more detailed analysis of the reasons behind the superior performance of the TATR model compared to the LGPMA model on the Vietnamese Tax Liability dataset but its inferior results on the Borderless Tables dataset, we surveyed the datasets utilized for the training of both models and the deep learning models employed by each. Firstly, the training dataset of the two models has been investigated. The TATR model was trained on a combined dataset from PubTables-1M and FinTabNet. The PubTables-1M contains nearly one million tables from science articles, while FinTabNet has approximately 113K financial report tables. The advanced data processing steps [3] were applied to both datasets; for example, annotation mistakes and inconsistencies were identified and fixed. For PubTabNet, the dataset used to train the LGPMA model is inferior both in terms of data quantity and precision in annotations. With respect to data quantity, PubTabNet contains only 568K tables, which is less than half of the combined dataset mentioned above. Regarding the consistency and precision in annotations, the annotations in the PubTabNet dataset have not been processed through high-level data processing steps. Therefore, the annotations in this dataset may lack consistency, and many columns and rows may not be fully and accurately annotated. The difference between the two training datasets is shown in Table 2.

Table 1. The scores on TEDS-Struc. metric

Testing Datasets	Models	TEDS-Struc. (%)
Borderless Tables	LGPMA	**96.54**
	TATR	89.15
Vietnamese Tax Liability Tables	LGPMA	52.83
	TATR	**88.98**

Table 2. Comparison of training datasets between the two models

Models	Training Dataset	Samples	Language	The Canonicalization Procedure
TATR	PubTables 1M + FinTabNet	1.06M	English	Yes
LGPMA	PubTabNet	510K	English	No

With a diverse dataset synthesized from scientific papers and financial reports, combined with rigorous verification in the data annotation step. This has contributed to the TATR model predicting outcomes more effectively than the LGPMA model when encountering previously unseen data, especially the Vietnamese Tax Liability dataset. Regarding the deep learning-based methods aspect, LGPMA was built from the Mask R-CNN network with ResNet-50 backbone; on the other hand, TATR was applied to the Detection Transformer (DETR) with Resnet-18 backbone. The weights of these two backbones were trained on two distinct datasets, namely MS-COCO and ImageNet, respectively.

LGPMA was divided into two tasks: Local Pyramid Mask Alignment (LPMA) for detecting aligned bounding boxes of cells and Global Pyramid Mask Alignment (GPMA) to handle empty cells. LPMA relies on character regions, while GPMA predicts both empty and non-empty cells by analyzing global table components. This approach improves performance on borderless tables but is sensitive to noise and low-quality images, such as scanned images. Figure 3 illustrates the prediction results of the LGPMA and TATR models using scanned image (Vietnamese Tax Liability Dataset) as input.

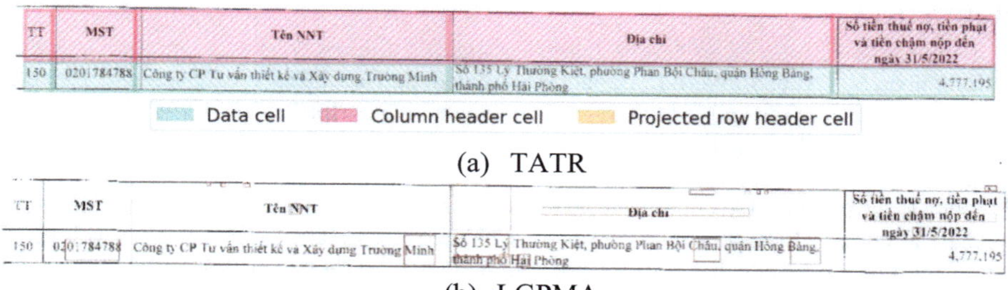

(a) TATR

(b) LGPMA

Fig. 3. The prediction results of TATR and LGPMA models on scanned image

Furthermore, both models are non-end-to-end approaches, only per forming the task of TSR without recognizing the content within each cell of the table. Therefore, the model size and inference time of these approaches should also be considered when integrating them into applications. Specifically, the size of the LGPMA model is approximately 280.5 Mb, while the TATR's size is 160.5 Mb. The LGPMA model has an inference time of 3.2 s per image, whereas TATR takes 5 s. Furthermore, installing and deploying the LGPMA model is more challenging than TATR due to strict dependencies between library versions in the environment requirements.

4 Conclusion

In this paper, we performed a comparative study between the two state-of-the art models through the years, namely LGPMA and TATR, using two datasets that we created to represent practical tables. The experiments showed that, with a diverse dataset synthesized from scientific papers and financial reports combined with rigorous verification in the data annotation step, the TATR model predicted outcomes more effectively than the LGPMA model when encountering unseen data, especially the Vietnamese Tax Liability dataset. In particular, the TATR model has an 88.98% score on the TEDS metric, while the LGPMA model only has a 52.83% score. However, with the Borderless Table dataset, the TATR model predicts the results less accurately than the LGPMA model, specifically scoring lower by 7.39%.

By comparing the inference results of these models, we realized that the practical applicability of these models depends not only on the algorithm but also on the diversity of table types and rigorous verification steps in creating of annotations for the training data. In the future, we anticipate creating a comprehensive Vietnamese Tax Liability dataset to facilitate training on LGPMA and TATR models.

Acknowledgements. We thank James Jackson for his fruitful discussions and comments on the topics of TEDS and LGPMA.

References

1. Schreiber, S., Agne, S., Wolf, I., Dengel, A., Ahmed, S.: DeepDeSRT: deep learning for detection and structure recognition of tables in document images. In: 2017 14th IAPR International Conference on Document Analysis and Recognition (ICDAR), vol. 1, pp. 1162–1167. IEEE (2017)
2. Qiao, L., et al.: LGPMA: complicated table structure recognition with local and global pyramid mask alignment. In: Lladós, J., Lopresti, D., Uchida, S. (eds.) ICDAR 2021. LNCS, vol. 12821, pp. 99–114. Springer, Cham (2021). https://doi.org/10.1007/978-3-030-86549-8_7
3. Smock, B., Pesala, R., Abraham, R.: PubTables-1M: towards comprehensive table extraction from unstructured documents. In: Proceedings of the IEEE/CVF Conference on Computer Vision and Pattern Recognition, pp. 4634–4642 (2022)
4. Kieninger, T., Dengel, A.: The T-Recs table recognition and analysis system. In: Lee, S.W., Nakano, Y. (eds.) DAS 1998. LNCS, vol. 1655, pp. 255–270. Springer, Heidelberg (1999). https://doi.org/10.1007/3-540-48172-9_21
5. Carion, N., Massa, F., Synnaeve, G., Usunier, N., Kirillov, A., Zagoruyko, S.: End-to-end object detection with transformers. In: Vedaldi, A., Bischof, H., Brox, T., Frahm, J.M. (eds.) ECCV 2020. LNCS, vol. 12346, pp. 213–229. Springer, Cham (2020). https://doi.org/10.1007/978-3-030-58452-8_13
6. Hurst, M.: A constraint-based approach to table structure derivation. In: Seventh International Conference on Document Analysis and Recognition (ICDAR), vol. 3, p. 911. IEEE Computer Society (2003)
7. Zhong, X., ShafieiBavani, E., Jimeno Yepes, A.: Image-based table recognition: data, model, and evaluation. In: Vedaldi, A., Bischof, H., Brox, T., Frahm, J.M. (eds.) ECCV 2020. LNCS, vol. 12366, pp. 564–580. Springer, Cham (2020). https://doi.org/10.1007/978-3-030-58589-1_34
8. Zheng, X., Burdick, D., Popa, L., Zhong, X., Wang, N.X.R.: Global table extractor (GTE): a framework for joint table identification and cell structure recognition using visual context. In: Proceedings of the IEEE/CVF Winter Conference on Applications of Computer Vision, pp. 697–706 (2021)

An Assessment of Formative and Reflective Constructs in Ethnic Consumer Behavior Research

Thi Thu Cuc Nguyen(✉) and Van Tang Nguyen

Foreign Trade University, Hanoi, Vietnam
{cucntt,tangnv}@ftu.edu.vn

Abstract. This study employs Confirmatory Tetrad Analysis (CTA) to evaluate the specification and validation of formative and reflective constructs within a Partial Least Squares Structural Equation Modeling (PLS-SEM) framework. While the formative-reflective distinction is critical in construct development, its implications are often overlooked, leading to issues such as construct misspecification, identification challenges, and validation errors. Using a systematic approach, this research compares the utility of formative and reflective modeling approaches, demonstrating that certain constructs align better with a reflective framework, while one other exhibit stronger validity when modeled formatively. The findings highlight the importance of adopting a hybrid formative-reflective approach to accurately capture the complexity of latent constructs and their observed measures.

Keywords: PLS-SEM · Formative-Reflective · Confirmatory tetrad analysis (CTA) · measurement model

1 Introduction

Despite its widespread adoption in social science research, Partial Least Squares Structural Equation Modeling (PLS-SEM) faces a critical challenge in the proper specification of measurement models, particularly regarding the distinction between reflective and formative constructs [1]. Indeed, the misclassification of these constructs can significantly compromise research validity through biased path coefficients and inflated measurement errors [2]. While reflective constructs operate on the principle that latent variables drive observable indicators - as evidenced in cases like customer satisfaction influencing purchasing behavior - formative constructs work in the opposite direction, with potentially uncorrelated indicators collectively defining the construct, such as in socioeconomic status measurements [3]. Unfortunately, despite their theoretical significance, researchers often default to reflective models simply due to their familiarity, overlooking the potential benefits of formative approaches [4]. The consequences of such misspecification are far-reaching, potentially leading to both Type I and Type II errors that can fundamentally distort theoretical conclusions [5]. Moreover, the problem compounds itself as researchers build upon previously mis-specified constructs, creating a cascade of methodological errors throughout literature.

J. S. Park et al. (Eds.): FutureTech 2025, LNEE 1483, pp. 31–37, 2025.
https://doi.org/10.1007/978-981-95-1999-6_5

This research therefore tackles these pressing issues head-on by developing a rigorous PLS-SEM framework for evaluating construct specifications. By providing clear guidelines for proper construct development and validation, while emphasizing the crucial role of theoretical considerations over mere methodological convenience, this study aims to enhance the reliability and validity of future SEM research in the social sciences.

2 Methods

2.1 The Model Structure

In this study, to represent the measurement models for latent variables, we use the following equations:

$$X = \Lambda_\xi \xi + \delta$$

$$Y = \Lambda_\eta \eta + \epsilon$$

where:

- X is a (p x 1) vector of observed indicators for exogenous variables.
- Y is a (q x 1) vector of observed indicators for endogenous variables.
- Λ_ξ is a (p x n) matrix of factor loadings for exogenous variables.
- Λ_η is a (q x m) matrix of factor loadings for endogenous variables.
- δ and ϵ are vectors of measurement errors for exogenous and endogenous variables, respectively.

2.2 The Instrument

This empirical investigation delves into the nature of Ethnic Identification (EID) within a PLS-SEM framework, specifically examining whether it manifests as a formative or reflective construct. Drawing from a diverse sample of Chinese (40%) and Vietnamese (60%) adults in the United Kingdom (n = 80), the study employs well-validated measurement scales across six key constructs. Specifically, the research utilizes cross-cultural validated scales, such as Keillor's NATID Scale, using a comprehensive 7-point Likert scale, providing robust data for the hypothesized relationships depicted in Fig. 1.

Note:

H1a: Nationalism has a direct negative effect on xenophilia among the UK's immigrants.

H2a: Consumer Ethnocentrism has a negative effect on xenophilia among the UK's immigrants.

H1b: The relationship between nationalism and xenophilia among the UK's immigrants will be mediated by ethnic identification.

H2b: The relationship between consumer ethnocentrism and xenophilia among the UK's immigrants will be mediated by ethnic identification.

H3a: The negative relationship between nationalism and xenophilia among the UK's immigrants will be stronger for Chinese when compared to Vietnamese.

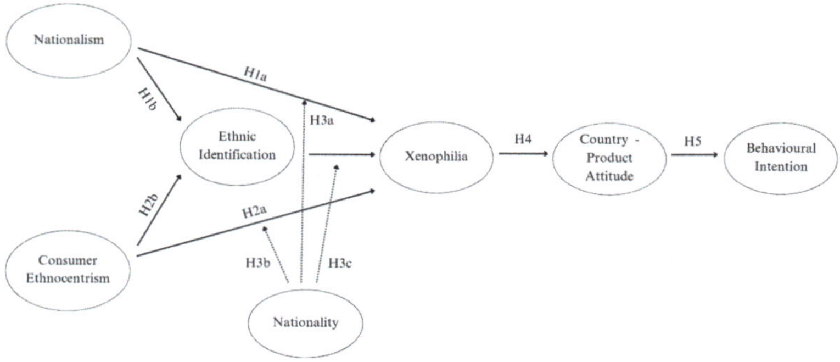

Fig. 1. Proposed Conceptual framework

H3b: The negative relationship between consumer ethnocentrism and xenophilia among the UK's immigrants will be stronger for Chinese when compared to Vietnamese.

H3c: The mediation effect of ethnic identification will be moderated by nationality among the UK's immigrants.

H4: Xenophilia has a positive effect on country-product attitude among the UK's immigrants.

H5: Country-product attitude has a positive effect on behavioural intention among the UK's immigrants.

3 Data Analysis and Result

3.1 Model Comparison

To assess the impact of construct specifications in Partial Least Squares Structural Equation Modeling (PLS-SEM), two structural models were compared. Model A specified all constructs, including Ethnic Identification (EID), as reflective, while Model B treated EID as a formative construct while maintaining the reflective specification for other constructs (Fig. 2). The models were estimated using SmartPLS 4.0, and path coefficients were analyzed to determine the effect of construct specification on hypothesis testing outcomes.

```
# Model A specification
reflective_measurement <- constructs(
  composite("NAT", multi_items("NAT", 1:3)),
  composite("CET", multi_items("CET", c(2,4,5,6))),
  composite("XEN", multi_items("XEN", 1:6)),
  composite("CPA", multi_items("CPA", 1:3)),
  composite("BEI", multi_items("BEI", 1:3)),
  composite("EID", multi_items("EID", 1:4)), # Reflective specification
  composite("NAT_Nationality", single_item("NAT_Nationality")),
  composite("CET_Nationality", single_item("CET_Nationality"))
)
```

```
# Model B specification
formative_measurement <- constructs(
  composite("NAT", multi_items("NAT", 1:3)),
  composite("CET", multi_items("CET", c(2,4,5,6))),
  composite("XEN", multi_items("XEN", 1:6)),
  composite("CPA", multi_items("CPA", 1:3)),
  composite("BEI", multi_items("BEI", 1:3)),
  composite("EID", multi_items("EID", 1:4), weights = mode_B), # Formative specification
  composite("NAT_Nationality", single_item("NAT_Nationality")),
  composite("CET_Nationality", single_item("CET_Nationality"))
)
```

Fig. 2. Differences in Specifications of Model A and Model B in R

As shown in Table 1, treating EID as a formative construct in Model B resulted in significant changes in key path relationships. Notably, the effect of EID on XEN

Table 1. Correlations and square root of average extracted of constructs

Path Relationship	Model A	Model B	Change
CET → EID	0.431	0.433	+0.002
CET → XEN	–0.603	–0.794	–0.191
CPA → BEI	0.69	0.702	+0.012
EID → XEN	1.213	1.541	+0.328 *
NAT → EID	0.437	0.432	–0.005
NAT → XEN	–0.639	–0.791	–0.152 *

* Significant changes observed (>0.15 absolute value difference).

Table 2. Comparison between Hypothesis Testing result

Hypothesis	Model A			Model B		
	Path Coefficient	t-value	Result	Path Coefficient	STDEV	Result
H1a	**-0.639**	1.150	Not supported	**-0.791**	1.172	Not supported
H1b	1.123	2.037	Not supported		2.029*	Supported
H2a	–0.603	1.911	Not supported	–0.794	2.050*	Supported
H2b	1.123	2.037	Not supported		1.844	Not supported
H3a	0.729	1.479	Not supported	0.912	1.929	Not supported
H3b	–1.135	1.895	Not supported	–1.465	1.897	Not supported
H3c	0.595	1.789	Not supported	0.790*	2.117	Supported
H4	0.691***	9.119	Supported	0.630***	9.113	Supported
H5	0.690***	10.451	Supported	0.702***	10.451	Supported

Note: * $p < 0.05$; ** $p < 0.01$; *** $p < 0.001$.

increased substantially ($\beta = 1.213$ in Model A vs. $\beta = 1.541$ in Model B), suggesting that a formative specification better captures EID's explanatory power. Similarly, the negative effects of CET and NAT on XEN became more pronounced underscoring the importance of construct specification in moderating relationships (Tables 2 and 3).

3.2 Confirmatory Tetrad Analysis (CTA)

To determine whether EID should be modeled as a formative or reflective construct, Confirmatory Tetrad Analysis (CTA) was conducted [6]. The results revealed that all tetrads failed to vanish, with 100% of them being statistically significant ($p < 0.05$), supporting the formative nature of EID. This finding reinforces that specifying EID as reflective, as in Model A, led to underestimation of its influence and misrepresentation of its structural role [7].

Table 3. Measurement Model Indices and Recommended Thresholds

Indices	Study Value	Recommended Value
Average R^2 (ARS)	0.412	
Average adjusted R^2 (AARS)	0.386	
Average block VIF (AVIF)	2.499	Acceptable if ≤ 5; ideally ≤ 3.3
SRMR Saturated	0.073	Acceptable if ≤ 0.08 (conservative) or ≤ 0.10
SRMR Estimated	0.089	
Cronbach's Alpha	0.766–0.891	Acceptable if ≥ 0.70
Composite Reliability (rho_a)	0.786–0.892	
Composite Reliability (rho_c)	0.864–0.932	
Average Variance Extracted (AVE)	0.624–0.821	Acceptable if ≥ 0.50
Outer Loadings	0.731–0.936	Acceptable if ≥ 0.70
VIF	1.00–2.118	Acceptable if ≤ 5.00
HTMT	0.096–0.839	Acceptable if ≤ 0.85

3.3 Structural Model Fit Analysis

The structural model comparison further highlights the necessity of correctly specifying formative constructs. Several hypotheses that were not supported in Model A due to low path significance (e.g., H1b, H2a, and H3c) became significant in Model B, reinforcing that construct specification influences hypothesis testing results. Notably, in Model B, the H1b ($\beta = 2.029$, $p < 0.05$), H2a ($\beta = -0.794$, $p < 0.05$), and H3c ($\beta = 0.790$, $p < 0.05$) became statistically significant when EID was modeled as formative, confirming that its nature affects its explanatory power within the structural model [8].

From a model fit perspective, Model B exhibits better theoretical consistency and statistical reliability, after bootstrapping resampling approach with 5,000 resamples. The significant paths in Model B reinforce the necessity of separating reflective and formative constructs to avoid constructing misspecification, which can lead to biased results. The CTA further supports the formative nature of EID, as all tetrads were statistically significant ($p < 0.05$), confirming that EID is best measured through its indicators rather than being influenced by them.

Model B exhibited satisfactory fit indices, with the smallest R-square value observed for XEN at 23% and the highest for EID at 52.8%. The SRMR value for the saturated model was 0.073, indicating good model fit [9]. The Normed Fit Index (NFI) of 0.735 suggested moderate fit [1], and the Chi-square value of 378.786 reflected a significant improvement over the baseline model. These results highlight the necessity of separating and examining formative constructs to avoid bias in parameter estimation, specifically EID, to improve model validity, allowing for a more precise understanding of the relationships among variables [10].

4 Conclusion

By incorporating EID as a formative construct, the model can better account for how immigrants' perceptions of their home and host countries influence their behavioral outcomes. For instance, strong ethnic identification may amplify or mitigate the adverse effects of nationalism and consumer ethnocentrism, depending on how immigrants navigate their dual identities. This leads to a deeper understanding of how immigrants' perceptions of their cultural roots and their adopted society can predict their behavioral tendencies, such as openness to foreign cultures or resistance to assimilation. Moreover, the Confirmatory Tetrad Analysis validated the hybrid approach of combining formative and reflective measurements, ensuring accurate construct representation. The findings highlight that overlooking formative constructs can lead to biased path coefficients and misinterpretations, reinforcing the need for theoretical and empirical justification in measurement model specification [11]. By adopting a hybrid model with EID as a formative construct, the study improved model validity and theoretical alignment, providing a more precise understanding of intervariable relationships. These insights contribute to advancing PLS-SEM applications in sociological and consumer behavior research.

References

1. Hair, J.F., et al.: Executing and interpreting applications of PLS-SEM: updates for family business researchers. J. Fam. Bus. Strat. **12**(3), 100392 (2021). https://doi.org/10.1016/j.jfbs.2020.100392
2. MacKenzie, S.B., Podsakoff, P.M., Jarvis, C.B.: The problem of measurement model misspecification in behavioral and organizational research and some recommended solutions. J. Appl. Psychol. **90**(4), 710–730 (2005). https://doi.org/10.1037/0021-9010.90.4.710
3. Diamantopoulos, A., Winklhofer, H.M.: Index construction with formative indicators: an alternative to scale development. J. Mark. Res. **38**(2), 269–277 (2001). https://doi.org/10.1509/jmkr.38.2.269.18845
4. Sarstedt, M., Hair, J.F., Ringle, C.M., Thiele, K.O., Gudergan, S.P.: Estimation issues with PLS and CBSEM: Where the bias lies! J. Bus. Res. **69**(10), 3998–4010 (2016). https://doi.org/10.1016/j.jbusres.2016.06.007
5. Ringle, C.M., Wende, S., Becker, J.-M.: SmartPLS 4." SmartPLS, Bönningstedt, 2024. https://www.smartpls.com/
6. org Henseler, J., Chin, W.W.: A comparison of approaches for the analysis of interaction effects between latent variables using partial least squares path modeling. Struct. Equation Model. Multidisciplinary J. **17**(1), 82–109, January 2010
7. Dybro Liengaard, B.: Measurement invariance testing in partial least squares structural equation modeling. J. Bus. Res. **177**, 114581, April 2024. https://doi.org/10.1016/j.jbusres.2024.114581
8. Henseler, J., Hubona, G., Ray, P.A.: Using PLS path modeling in new technology research: updated guidelines. Ind. Manag. Data Syst. **116**(1), 2–20 (2016). https://doi.org/10.1108/IMDS-09-2015-0382
9. Henseler, J., Ringle, C.M., Sarstedt, M.: Testing measurement invariance of composites using partial least squares. Int. Mark. Rev. **33**(3), 405–431 (2016). https://doi.org/10.1108/IMR-09-2014-0304

10. Wang, S., Cheah, J.-H., Wong, C.Y., Ramayah, T.: Progress in partial least squares structural equation modeling use in logistics and supply chain management in the last decade: a structured literature review. IJPDLM **54**(7/8), 673–704 (2024). https://doi.org/10.1108/IJPDLM-06-2023-0200
11. Kante, M., Michel, B.: Use of partial least squares structural equation modelling (PLS-SEM) in privacy and disclosure research on social network sites: a systematic review. Comput. Hum. Behav. Rep. **10**, 100291 (2023). https://doi.org/10.1016/j.chbr.2023.100291

Pothole Detection and Classification Using YOLO Models

Thien Thanh Tran Ngoc[1], Tien Thanh Do[1(✉)], Ngoc Anh Le[2], Thi Nhung Vuong[3], Ngoc Thanh Pham[1], Doan Dong Nguyen[1], and Phuong Anh Nguyen[2]

[1] International School, Viet Nam National University, 144 Xuan Thuy Street, Cau Giay District, Hanoi 10000, Viet Nam
thanhdo@vnuis.edu.vn

[2] Swinburne University of Technology (Vietnam), 2 Duong Khue Street, Mai Dich Ward, Hanoi 10000, Viet Nam

[3] Hanoi School of Business and Management, Viet Nam National University, 144 Xuan Thuy Street, Cau Giay District, Hanoi 10000, Viet Nam

Abstract. The study introduces an advanced approach to pothole detection and classification implemented by YOLOv12 and shows greater performance improvements than YOLOv8. A new data set was created containing 1,000 pothole images (water and no water). The images, along with staggered ground truths, were recorded manually. We ran five different models on this data merged from YOLOv8 and YOLOv12. While, YOLOv12 surpasses YOLOv8 in detection and generalization, YOLOv8 outperformed YOLOv12 in classifying ponds. These results establish a foundation for integrating YOLO-based pothole detection into intelligent transportation systems, enhancing road safety and infrastructure maintenance. Future research may focus on further architectural optimizations and real-world deployment strategies to enhance system efficiency.

Keywords: YOLO · Pothole Detection · Intelligent Transportation Systems · Computer vision

1 Introduction

Roads play a vital role of modern life helping people and goods move around quickly and safely. However, potholes are a common problem that can make roads unsafe. Potholes are holes or cracks that appear on the road surface, usually because of water damage or heavy traffic. Potholes are holes or cracks that arise on the road surface under the action of water and traffic. They throw riders off balance, particularly those on motorbikes and bicycles, who often come to grief after having hit one [1]. In 2021, potholes accounted for 0.8% of road accidents, 1.4% of traffic fatalities and 0.6% injuries [2].

Today, most potholes are found by workers who check roads manually. This working pattern not only has its problems in terms of expenditure but also can result in consistency of results with each worker [3]. Even where it is used at present, a good many systems of technology still require human input or are only semi-automatic [4, 5]. In recent years,

J. S. Park et al. (Eds.): FutureTech 2025, LNEE 1483, pp. 38–48, 2025.
https://doi.org/10.1007/978-981-95-1999-6_6

new measures using computer vision and deep learning have been developed to combat this problem. Among these, one of the most popular is called the YOLO (You Only Look Once) model which accurately and rapidly detects objects in images [6]. Fast and adaptable for many different situations, YOLO has become very common [7].

But there are so many different versions of YOLO nowadays and each has its own strengths and weaknesses. Especially when it comes to small devices such as drones or smart cars, it takes great care to find out which one works best [8].

This study presents a comprehensive comparison between two recent object detection architectures—YOLOv8 and YOLOv12—for pothole detection and classification under real-world conditions. The main contributions are summarized as follows:

- A new manually annotated dataset is introduced, containing 1,000 high-resolution images of potholes in various conditions (with and without water), along with diverse augmentation techniques to improve generalization.
- Five model variants from each architecture (YOLOv8 and YOLOv12) are benchmarked, with evaluations concentrating on detection accuracy, classification performance, and computational efficiency.
- Lightweight models YOLOv8n and YOLOv12n are analyzed to assess their suitability for real-time and resource-constrained applications, such as deployment on embedded systems and UAVs.

2 Material and Method

2.1 Dataset

In this study, we collected pictures of potholes in different states, some with water and some without. We filter out any blurry images that did not meet a certain quality standard, ensuring that only high-quality pictures were kept and carefully annotated, as shown in Fig. 1. This led to a dataset of carefully selected images, with 1000 annotated pictures; which we call the "POTHOLES" dataset. The training dataset, used for model training, represented 85% of the overall data; the validation and test sets were represented, respectively, 5–10%. More information on the dataset is also presented in Table 1.

Furthermore, we adopted different image processing techniques to improve our data set: turning left and right, up and down, increasing brightness to 25% while decreasing it by 25%, lowering exposure levels by 15% to increase them with roughly 15%, combining a reduction in clarity and other operations. All images are re-sized uniformly into 640 × 640 format (Fig. 2).

Table 1. Number of images in the datasets

Category	Training	Validation	Testing	Total
Image	792	197	11	1000
Augmentation	2080	197	11	2379

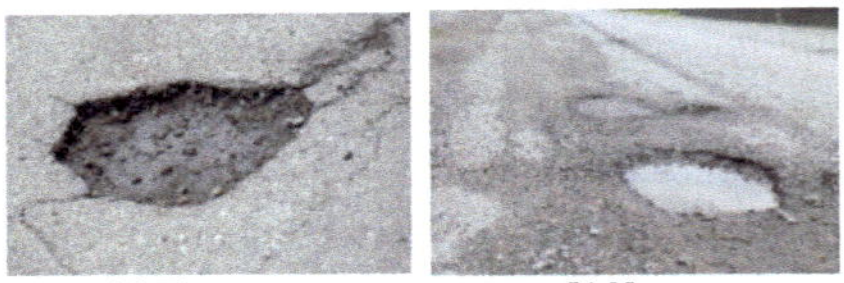

(a) Water (b) No water

Fig. 1. Examples of potholes for two examples

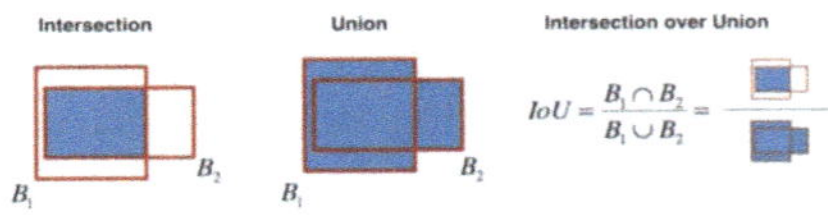

Fig. 2. Intersection over Union [8]

2.2 Metrics

The mathematical expressions for precision and recall are given by:

$$\begin{cases} \text{Precision} = \frac{\text{TP}}{\text{TP+FP}} \\ \text{Recall} = \frac{\text{TP}}{\text{TP+FN}} \end{cases} \tag{1}$$

In this context, TP (True Positives), FP (False Positives) and FN (False Negatives) denote the number of correctly detected objects, incorrectly detected objects, and missed detections, respectively. Qualitative error analysis is an essential step when comparing the performance of YOLOv8 and YOLOv12. While quantitative metrics provide an overall view, visually examining false positive and false negative cases helps to reveal the biases, error patterns, and specific limitations of each model. This information is invaluable for selecting and optimizing a model for deployment in applications that demand high safety and accuracy.

The calculation of Average Precision (AP) and mean Average Precision (mAP) is utilized to quantitatively evaluate the performance of the object detection model:

$$\begin{cases} AP = \int_0^1 \text{Precision}(\text{Recall})d\text{Recall} \\ \text{mAP} = \frac{1}{\text{N}} \sum_{\text{i}=1}^{\text{N}} \text{AP}_{\text{i}} \end{cases} \tag{2}$$

2.3 YOLO Series

YOLO architecture

You Only Look Once (YOLO) [1] was created to solve the single-pass object detection problem. In 2015, Joseph Redmon and his team launched it as an alternative to traditional approaches (e.g., R-CNN and Faster R-CNN). For object detection in computer vision, YOLO uses the single convolutional neural network (CNN) (Fig. 3) to process an entire image that is then broken down into a grid structure. This enables a much higher statistical accuracy [2] and significantly increases the processing precision. For each grid cell, the model predicts bounding boxes along with their corresponding confidence scores; these, in turn, are used to gauge how likely it is that there is an object inside this box. The boundaries of these bounding boxes are then further refined using predicted confidence scores.

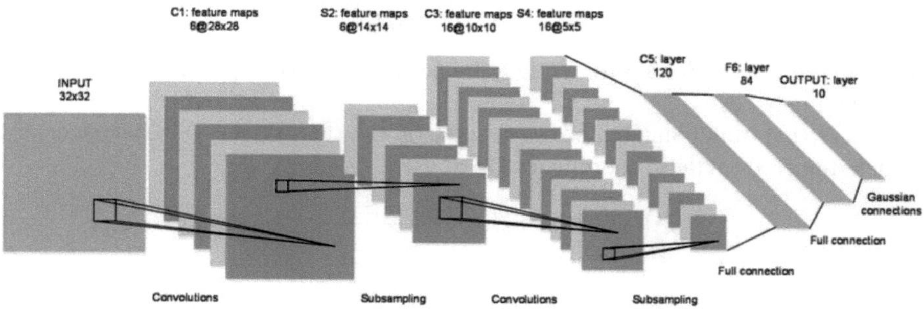

Fig. 3. The structure of CNN [3]

The YOLO architecture consists of 24 convolutional layers followed by two fully connected layers [4], allowing effective real-time object detection. A visual representation of the architecture can be found in Fig. 4. YOLO takes a 448 x 448 pixel image as input (after transforming the data, this would be called "preprocessing") and reshapes it into the square required by the initial input layer. Moving through convolutional neural networks, it is eventually transformed into a (7 x 7 x 30) tensor. This tensor stores all the information necessary for object detection: specifically, it has coordinates for bounding boxes, plus a distribution over all classes on which our model was trained. A thresholding mechanism processes confidence scores before units are passed into the next layer.

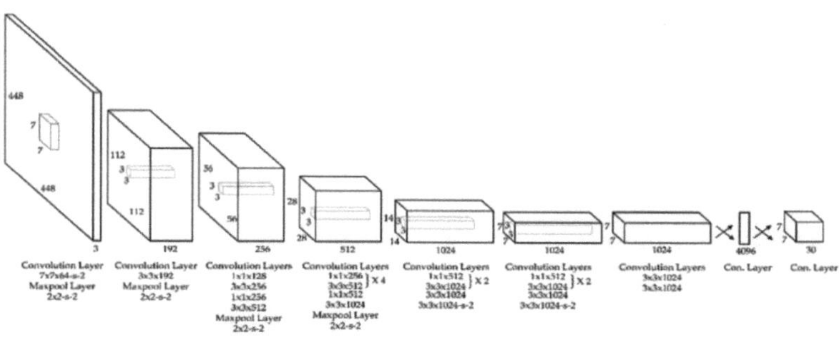

Fig. 4. The structure of YOLO [5]

Leaky Rectified Linear Unit

Leaky RELU is a novel variant of the Rectified Linear Unit (ReLU) that addresses the problem of vanishing gradients when the input takes negative value [1]:

$$\left\{\varphi(\mathrm{x}) = \begin{array}{ll} x, & \text{if x} > 0 \\ 0.1x, & \text{otherwise} \end{array}\right. \tag{3}$$

For x > 0, the fuction return value x itself. For $x \leq 0$, instead of being set to 0 as in traditional ReLU, Leaky ReLU allows a small fraction of negative values to pass through, thus preventing neuron death and improving the model's learning capacity.

Bounding box
In the architecture proposed for YOLO, the neck module is responsible for predicting the bounding boxes.

The ground truth bounding box is denoted by $G = (x_1, y_1, x_2, y_2)$. The edges of G are determined from these coordinates [13]:

$$\begin{cases} tx_1 = \log \frac{s_l(x+0.5)-x_1}{r_l} \\ tx_1 = \log \frac{s_l(y+0.5)-y_1}{r_l} \\ tx_1 = \log \frac{x_{2-s_l}(x+0.5)}{r_l} \\ tx_1 = \log \frac{y_{2-s_l}(y+0.5)}{r_l} \end{cases} \tag{4}$$

Loss function

$$\mathcal{L} = \lambda_{coord} \sum_{i=0}^{S^2} \sum_{j=0}^{B} 1_{ij}^{obj} \left[(x_{1-}\hat{x}_i)^2 + (y_{1-}\hat{y}_i)^2 \right] + \lambda_{coord} \sum_{i=0}^{S^2} \sum_{j=0}^{B} 1_{ij}^{obj} \left[\left(\sqrt{w_i} - \sqrt{\hat{w}_i}\right)^2 + \left(\sqrt{h_i} - \sqrt{\hat{h}_i}\right)^2 \right] + \sum_{i=0}^{S^2} \sum_{j=0}^{B} 1_{ij}^{obj} (C_{1-}\hat{C}_i)^2 + \lambda_{noobj} \sum_{i=0}^{S^2} \sum_{j=0}^{B} 1_{ij}^{noobj} (C_{1-}\hat{C}_i)^2 + \sum_{i=0}^{S^2} 1_{ij}^{obj} \sum_{cclasses} (p_i(c) - \hat{p}_i(c))^2 \ [6] \tag{5}$$

where 1_{ij}^{obj} indicates the presence of an object in cell i, while 1_{ij}^{obj} means that the jth bounding box predictor cell I is responsible for making the prediction.

3 Experimental Result

3.1 YOLOv8

YOLOv8 Model
YOLOv8 employs a structured design to optimize feature extraction and object detection efficiency. In YOLOv8, there are three test components. They are the backbone, neck, and head. The Backbone employs the C2f block in four stages (blocks: 2, 4, 6, 8), adding shortcut connections to enhance gradient flow and training stability. The depth-multiple parameters specifies how many bottleneck layers there are in C2f, while width-multiple max- channels control output channels. Two convolutional layers with size 3 and stride 2 make up the stem, downsizing input resolution and extracting initial feature representations. Down- sampling takes place by means of a convolutional block with step size 2 that halves spatial resolution. The stride also determines the spatial resolution of the down- sampling process and the amount by which to decrease the image size in each stage when moving from left to right.

The Neck contains four additional stages (blocks: 12, 15, 18, 21), and shortcut connections are eliminated based on empirical optimization. The SPPF (Spatial Pyramid Pooling Fast) block in the neck, succeeding the back-bone, provides multi-scale features; And it quickly uses pooling at different scales to increase feature information content outputted by the last layer. In addition, it merges features using concatenation and nearest-neighbor up-sampling. Concatenation aggregates different channels into one feature map with the same spatial resolution, while up-sampling increases the map resolution by replicating the values of adjacent pixels.

The Head is in charge of multinomial object recognition. It includes three detection heads. The first head is optimized detection for small objects and is thus placed on block 15; the second head, at block 18, picks out medium items; whereas the third head, associated with block 21, calls attention to large objects. The size of objects for YOLOv8 is relative to the dimensions of the image or video frame and is guaranteed across all YOLO versions. This design combines both computation and computation trading off efficiency against accuracy, as it provides a modular method for extracting features effectively and performing multiscale object detection (Table 2 and Fig. 5).

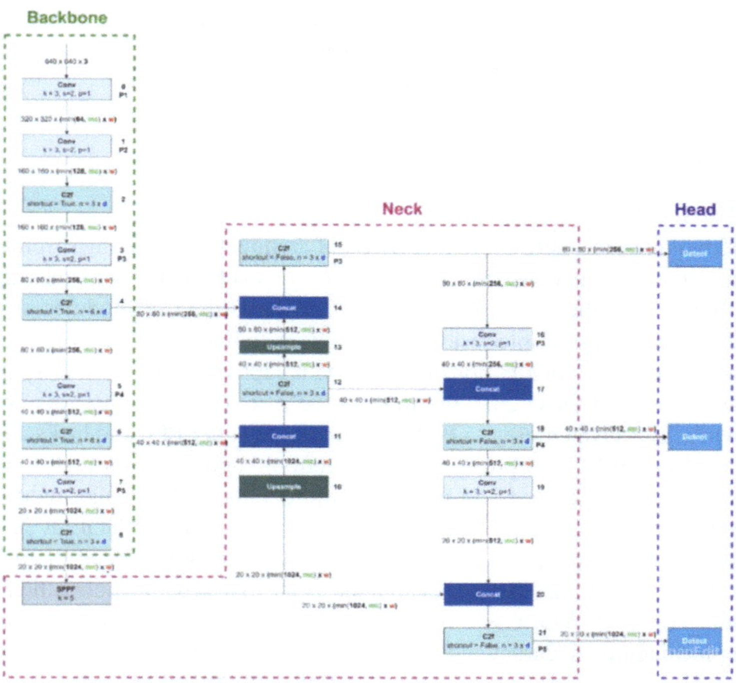

Fig. 5. Structure of YOLOv8 [7]

YOLOv8 Results

We used five variants of the YOLOv8.pt model to train and evaluate it on pothole data, via Google Colab, with High-RAM Colab Runtime and a T4 GPU. Tables 3 and 4 respectively detail the performance of these model variations.

Table 2. Comparision of Yolov12 variations of detection

Model	Params	P	R	mAP50	mAP50-90
Yolov8n	3.0	0.869	0.807	0.85	0.566
Yolov8s	9.4	0.886	0.799	0.863	0.603
Yolov8m	20.0	0.924	0.759	0.827	0.581
Yolov8l	43.6	0.904	0.781	0.843	0.583
Yolov8x	68.1	0.911	0.772	0.853	0.59

Table 3. Yolov8 results of pothole classification

Category	Instance	P	R	mAP50	mAP50–90
Water	213	0.794	0.688	0.751	0.423
No water	232	0.623	0.585	0.606	0.365
All	445	0.709	0.636	0.678	0.394

3.2 YOLOv12

YOLOv12 Model

YOLOv12 builds on the YOLOv8 and YOLOv11 architectures, making several architectural enhancements to increase the level of accuracy and efficiency while also improving its adaptability. With the Backbone, YOLOv12 replaces the C2f modules of YOLOv8 with R-ELAN, adding multiple residual connections to improve gradient flow. The 7 $\times$ 7 depth wise separable convolutions expand the receptive field, preserve spatial information, and reduce FLOPs compared to standard 3 $\times$ 3 kernels. In terms of the neck portion, while YOLOv8 employs a PANet architecture with SPPF and lacks an attention mechanism, YOLOv12 incorporates Area Attention—optimized through Flash Attention—which enables the model to focus on key regions in the image, reducing noise and maintaining high inference speed. This optimization allows YOLOv12 to obtain better expression efficiency for real-time detection tasks. In the Head, both YOLOv8 and YOLOv12 use an anchor-free, decoupled head that separates classification from localization. However, YOLOv12 refines the feature flow and loss function, thereby improving localization accuracy and the overall quality of the bounding boxes.

YOLOv12 typically achieves a higher mAP than YOLOv8 and offers better small-object detection. Through the use of depth wise separable convolutions and Flash Attention, YOLOv12 not only preserves or even improves inference speed compared to YOLOv8 but also consumes fewer parameters. Area Attention further enables the model to handle small or complex scenes effectively, reducing false positives in crowded or noisy environments and enhancing bounding box quality for tracking. In addition, YOLOv12's efficient resource usage (memory and parameter count) makes it highly suitable for embedded devices or drones where real-time performance is essential (Fig. 6).

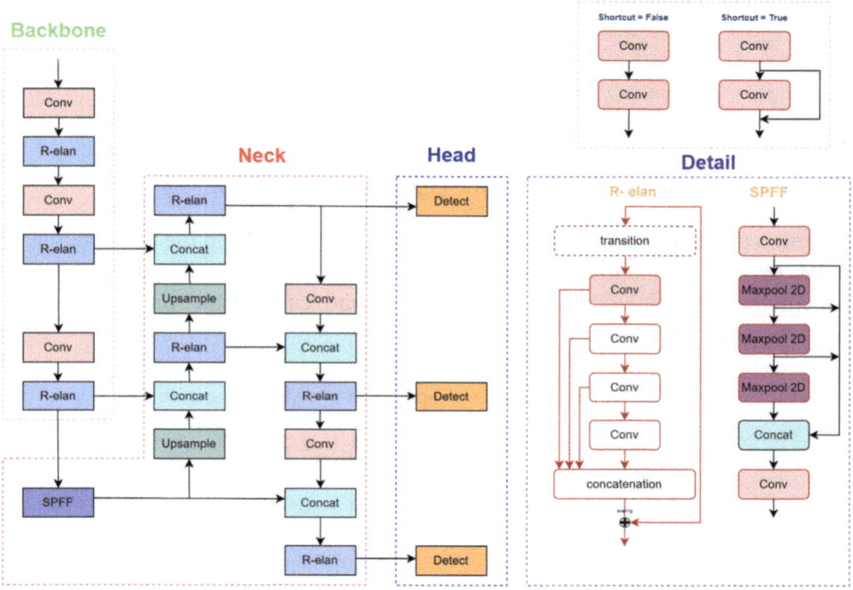

Fig. 6. Structure of YOLOv12

YOLOv12 Results

In this part, we conducted a full training course and evaluated five variations of the YOLO 12.pt model for the pothole dataset using Google Colab. Logically dividing the various models, each of them has been compared with the results of Tables 4 and 5.

Table 4. Comparision of Yolov12 variations of detection

Model	Params	P	R	mAP50	mAP50–90
Yolov12n	2.6	0.903	0.782	0.862	0.595
Yolov12s	9.4	0.886	0.799	0.863	0.603
Yolov12m	20.0	0.924	0.759	0.827	0.581
Yolov12l	25.3	0.902	0.766	0.838	0.603
Yolov12x	56.8	0.895	0.792	0.839	0.586

Table 5. Yolov12 result of pothole classification

Category	Instance	P	R	mAP50	mAP50–90
Water	213	0.876	0.665	0.771	0.471
No water	232	0.793	0.529	0.634	0.398
All	445	0.835	0.597	0.703	0. 434

3.3 Comparison of Version on a New Dataset

This section presents a detailed comparison of YOLO models on the newly curated pothole dataset. The evaluation focuses on detection accuracy, classification performance, and computational efficiency. Tables 6 and 7 summarizes the detection and classification performance of YOLOv8 and YOLOv12.

Table 6. Performance comparison for pothole detection

Model	P	R	mAP50	mAP50–90
Yolov8n	0.869	0.807	0.85	0.566
Yolov12n	0.903	0.782	0.862	0.595

Table 7. Performance comparison for pothole classification

Model	P	R	mAP50	mAP50–90
Yolov8n	0.869	0.807	0.85	0.566
Yolov12n	0.903	0.782	0.862	0.595

To compare the performance of YOLOv8 and YOLOv12 in real-world conditions, both models were tested on new dataset. This evaluation focused on detection accuracy, classification ability (water-filled or dry potholes), and efficiency when running on limited hardware. The key results are presented in Tables 6 and 7, with an emphasis on the lightweight models YOLOv8n and YOLOv12n.

For pothole detection, YOLOv12n showed slightly better performance than YOLOv8n. As reported in Table 7, YOLOv12n achieved a higher precision (0.903 vs. 0.869), and better mAP@50 (0.862 vs. 0.85) and mAP@50–90 (0.595 vs. 0.566). Although YOLOv8n had a marginally better recall, YOLOv12n delivered more consistent results overall. This improvement is likely due to YOLOv12's enhanced architecture, including the R-ELAN backbone, 7 × 7 depth-wise separable convolutions, and Flash Attention, which enable the model to better detect potholes under various conditions such as complex backgrounds or small object sizes [15].

In contrast, for classification (water-filled vs. dry potholes), YOLOv8n performed better than YOLOv12n. According to Table 7, YOLOv8n achieved higher recall and slightly better classification accuracy for the water-filled class. This advantage may stem from YOLOv8's more balanced and efficient architecture. Its anchor-free design and decoupled detection head allow for better generalization, especially when working with limited training data [16]. YOLOv12, although stronger in localization, may focus too heavily on bounding box quality, which can reduce sensitivity to fine-grained visual differences like water reflection or texture [17].

Both YOLOv8n and YOLOv12n are lightweight and suitable for deployment in real-time systems such as drones, embedded cameras, or intelligent transportation platforms. While YOLOv12n is preferable for precise object detection and localization, YOLOv8n remains a practical choice for classification tasks on small datasets, where simpler architectures may generalize better.

In summary, YOLOv12n is more effective for pothole detection, especially in complex scenes, whereas YOLOv8n performs better in classifying pothole types, making it a strong option when computational simplicity and classification reliability are the priorities (Figs. 7 and 8).

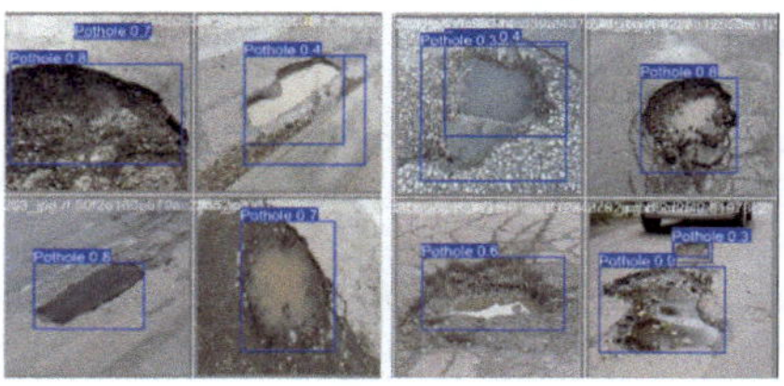

Fig. 7. Example of Pothole detection

Fig. 8. Example of Pothole classification

4 Conclusion

To conclude, this research illustrates that compared to YOLOv8, YOLOv12 has substantially higher detection accuracy and generalization performance. Using a thousand pothole images and manually classifying them, we systematically tested different versions of YOLO. This study finds that YOLOv12 has better accuracy than YOLOv8 in terms of accuracy, computing efficiency, and scale independence. Furthermore, with a balance between detection accuracy and processing speed, YOLOv12n and YOLOv8n emerge as real-time solutions. These light-weight models, in particular, are more suitable for environments which are resource restricted such as embedded systems, drones (unmanned aerial vehicles) or smart transportation systems. For the next stage, it is hoped that subsequent research will involve structural optimization of the model itself and a fine-tuning deployment strategy to provide a YOLO-based pothole detection system that really works in real-world maintenance of civil engineering structure.

References

1. Xing, C., et al.: A lightweight detection method of pavement potholes based on binocular stereo vision and deep learning. In: Proceedings of the International Conference on Computer Vision Systems (ICCVS 2024), p. 136733. Springer, Cham (2024)
2. Ali, F., et al.: Evaluating the effect of road surface potholes using a microscopic traffic model. J. Traffic Transp. Eng. **13**(15), 8677 (2023)
3. Fan, R., Tsui, M.: Road damage detection based on unsupervised disparity map segmentation. J. Transp. Res. **21**(11), 4906–4911 (2019)
4. Koch, C., et al.: A review on computer vision-based defect detection and condition assessment of concrete and asphalt civil infrastructure. Autom. Constr. **29**(2), 196–210 (2015)
5. Park, S.S., Tran, V.T., Lee, D.E.: Application of various YOLO models for computer vision-based real-time pothole detection. Sensors **11**(23), 11229 (2021)
6. Zou, Z., et al.: Object detection in 20 years: a survey. IEEE Trans. Pattern Anal. Mach. Intell. **111**(3), 257–276 (2023)
7. Yin, Y., Li, H., Fu, J.: Faster R-CNN: an accurate and faster object detection method. Comput. Vis. Patt. Recogn. **102**, 102756 (2022)
8. Redmon, J., et al.: You only look once: unified, real-time object detection. In: Proceedings of the IEEE Conference on Computer Vision and Pattern Recognition (CVPR 2016), pp. 1–9. Springer, Cham (2016)
9. Pan, Y., Wang, G., Yu, J.: Overview of deep learning YOLO algorithm. In: Proceedings of the Fourth International Conference on Computer Vision, Application, and Algorithm (CVAA 2024), p. 2025. Springer, Cham (2025)
10. Tra, V., et al.: Bearing fault diagnosis under variable speed using convolutional neural networks and the stochastic diagonal Levenberg-Marquardt algorithm. Mech. Syst. Signal Process. **17**(12), 2834 (2017)
11. Chablan, M.J.: YOLOv6: you only look once, real-time object detection explained. Technical Report, p. 21 (2017)
12. Koylu, C., Zhao, C., Shao, J.: Deep neural networks and kernel density estimation for detecting human activity patterns from geo-tagged images: a case study of birdwatching on Flickr. ISPRS J. Photogramm. Remote Sens. **81**(1), 45 (2019)
13. Setyawati, A., Kusdianto, I.J.P.E.: The effect of pavement condition on vehicle speeds and motor vehicles emissions. Transp. Res. Proc. **125**, 424–430 (2015)
14. Hidayatullah, P., et al.: YOLOv8 to YOLO11: a comprehensive architecture in-depth comparative review. In: Proceedings of the International Conference on Intelligent Vision Systems (ICIVS 2025), pp. 1–10. Springer, Cham (2025)
15. Al Rabbani Alif, M., Hussain, M.J.A.: YOLOv12: a breakdown of the key architectural features. arXiv preprint arXiv:2502.14740 (2025)
16. Terven, J., et al.: A comprehensive review of YOLO architectures in computer vision: from YOLOv1 to YOLOv8 and YOLO-NAS. Artif. Intell. Rev. **54**(3), 1680–1716 (2023)
17. Jegham, N., et al.: YOLO evolution: a comprehensive benchmark and architectural review of YOLOv12, YOLOv11, and their previous versions. In: Proceedings of the 2025 International Conference on Vision Engineering and Applications (VEA 2025), pp. 1–12. Springer, Cham (2025)

An Efficient Data Auditing Scheme for Multi-cloud Multi-replica Environments

Haiyan Yu, Yuxin Cui, and Chen Wang(✉)

Zhejiang Sci-Tech University, Hangzhou 310018, China
yuhaiyan0420@126.com, cuiyuxin_uni@163.com, wangchen@zstu.edu.cn

Abstract. With the rapid development of information technology, cloud computing has become increasingly popular, attracting a large number of users to outsource personal or corporate data to the cloud for storage and management. However, with the growth of data outsourced to the cloud, issues of data integrity and security have become increasingly prominent. These issues have become a focus of concern for both users and cloud service providers. We propose a novel auditing scheme for multi-cloud multi-replica storage. This scheme leverages a cloud combiner for coordination. Additionally, we design a bit-map-based file storage mapping mechanism. This scheme can audit multiple replicas of files stored on different cloud servers and provide feedback on the integrity and storage locations of the corresponding replica files. Therefore, the proposed scheme can enhance data security and auditing efficiency in a multi-cloud environment. Security analysis and performance evaluation both indicate that the scheme provides reliable and efficient auditing for outsourced data in a multi-cloud setting.

Keywords: Auditing · Multi-replica Storage · Multi-cloud

1 Introduction

In the context of the rapid development of information technology, cloud computing has emerged as a key driving force. It has greatly advanced modern information processing and storage technologies, bringing unprecedented convenience to data storage, processing, and sharing. Cloud storage, as an important subsystem of cloud computing, can offer powerful data capacity [1]. With cloud storage services, users can upload their data to the cloud to overcome the limitations of local storage capacity. However, when users outsource their data to the cloud instead of storing it locally, they lose physical control over their data. Since the cloud is susceptible to both internal and external security attacks, there is a risk of data loss or corruption [2]. Moreover, to preserve its reputation and avoid economic losses, a cloud service provider (CSP) may lie to users about data loss [3]. Therefore, ensuring the integrity of remotely outsourced data has become an issue that cannot be ignored when choosing cloud storage services.

Cloud auditing technology has emerged to enable remote verification of data integrity. However, in the information age, the explosive growth of data volumes and increasingly complex storage requirements pose new challenges. Single-file storage or

J. S. Park et al. (Eds.): FutureTech 2025, LNEE 1483, pp. 49–57, 2025.
https://doi.org/10.1007/978-981-95-1999-6_7

reliance on a single cloud server both pose risks: if a file is damaged or a cloud server fails, the data may become irrecoverable or lost. Therefore, users prefer to distribute their data across multiple cloud servers and create multiple copies to enhance data security and reliability. Existing audit solutions mostly introduce a cloud combiner to assist in the auditing.

Motivation of This Paper: However, the practice of users storing data across multiple CSPs complicates data management and security auditing. Meanwhile, privacy issues arising from data outsourcing cannot be ignored. Therefore, how to achieve secure and efficient data auditing in a multi-cloud environment remains a problem that requires further research.

In summary, our contributions are listed as follows:

A novel multi-cloud collaborative auditing scheme coordinated by a cloud combiner is proposed, which can ensure the integrity of all replica files uploaded and stored on different cloud servers.

A bit-map-based file storage mapping mechanism for multiple replica files has been proposed. It provides low storage overhead, accurate representation of file distribution across multiple clouds, and supports fast querying and flexible scalability.

By using pseudorandom permutation and hash function, the scheme stops the Third-Party Auditor (TPA) from inferring user privacy via audit responses, ensuring data security and privacy in a multi-cloud environment.

2 Related Work

In cloud storage, verifying the integrity of remote data is a crucial security issue. Malicious CSPs, driven by profit or other motives, may tamper with or corrupt the data. This poses a significant threat to data integrity. To check the integrity of outsourced data, Ateniese et al. [4] first proposed the Provable Data Possession (PDP) scheme in 2007. This scheme allows clients to verify the integrity of data stored on untrusted cloud servers and avoids the need to down-load the entire dataset. Juels et al. [5] defined and explored Proofs of Retrievability (PoR), in which users can verify both file integrity and data retrievability. Subsequently, cloud auditing has been extensively investigated, and many significant schemes have been proposed.

In practical applications, privacy protection is an important issue in data auditing. To address the issue of privacy protection in data auditing, researchers have proposed a variety of solutions. In 2010, Wang et al. [6] combined the public key based homomorphic authenticator with random masking and achieved the privacy-preserving public cloud data auditing system. Since then, an increasing number of auditing schemes based on privacy protection have been proposed [7–9].

However, these schemes typically store a single file in the cloud. Which can't be recovered if lost or damaged. So, storing multiple replicas has become an effective way to avoid such issues. In multi-replica storage, a file can be recovered as long as one copy is intact. The first multi-replica auditing scheme was proposed by Curtmola et al. [10] in 2008. This work extended proofs of data possession from single-file replicas to multi-replica scenarios in cloud storage systems. Subsequently, many multi-replica data auditing schemes in cloud storage were proposed [11, 12]. The aforementioned schemes store all replicas on one cloud service. However, storing data with a single CSP also carries risks; if the CSP is compromised, the data may be compromised as well. Moreover, CSPs might be curious about the data content or engage in data mining, leading to privacy breaches. In 2015, Wang et al. [13] proposed an identity-based distributed Provable Data Possession (PDP) scheme for multi-cloud storage. In 2021, Cai et al. [14] further revised and improved the security vulnerabilities of this scheme.

To summarize, with the increasing popularity of cloud computing, researchers have extensively explored multi-cloud and multi-replica data auditing. However, ensuring the integrity of each replica in a multi-cloud environment while protecting privacy still requires further research and improvement.

3 Preliminaries

Some important preliminaries, including Bilinear pairing, and discrete logarithm problem, for this paper are listed in this section.

3.1 Bilinear Pairing

Let q be a large prime number, G_1 and G_2 be two groups of order q, with operations defined as addition and multiplication respectively. The bilinear map e: $G_1 \times G_1 \rightarrow G_2$ satisfies the following properties:

(1) Bilinearity: For any $g_1, g_2 \in G_1$ and $a, b \in Z_q^*$, it holds that $e(g_1^a, g_2^b) = e(g_1, g_2)^{ab}$.
(2) Non-degeneracy: There exist $g_1 \in G_1$, such that $e(g_1, g_1) \neq 1$.
(3) Computability: For any $P, Q \in G_1$, there exists an efficient algorithm to compute $e(P, Q)$.

3.2 Discrete Logarithm Problem

Given $(g, g^a) \in G_1$, $a \in Z_q^*$, discrete logarithm (DL) problem is to calculate a. If the likelihood of an adversary solving the DL problem in the group G_1 in any polynomial time is negligible, then the DL problem cannot be computed in the group G_1.

4 Scheme Model

The system model mainly includes five entities: Cloud Server, Cloud Combiner, Key Generation Center (KGC), Third-Party Auditor (TPA), and User, as shown in Fig. 1. The specific description is as follows:

(1) Cloud Server: The cloud server is semi-honest and will still correctly execute the algorithms set by the system. It is responsible for storing the files uploaded by users. With significant storage space and computing resources, the cloud provides an auditing proof to verify whether files in the cloud are stored correctly and completely.

(2) Cloud Combiner: The cloud combiner is a trusted third party. It stores information on the placement of uploaded files, accepts challenges from the TPA, forwards the challenges to the corresponding cloud servers, aggregates the responses, and returns the results to the TPA.
(3) Key Generation Center (KGC): The honest and trusted KGC generates keys based on user identities and securely communicates with users through a secure channel to prevent key leakage or misuse.
(4) User: The user is an entity that possesses a large amount of data and wishes to store it on multiple cloud servers. The user can delegate the TPA to issue challenges to the cloud combiner to audit the integrity of all files stored on the cloud servers.
(5) TPA: The TPA will honestly and correctly execute the system's algorithms, but it is curious. Upon receiving an auditing request, the TPA initiates a challenge to the cloud combiner to verify the integrity of all files stored in the cloud.

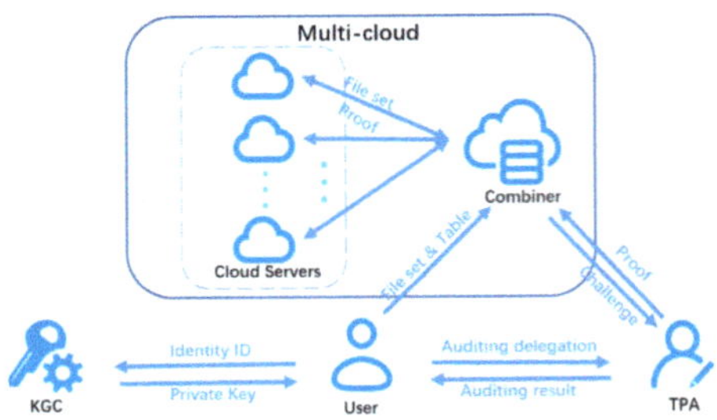

Fig. 1. System model diagram

5 Our Proposed Scheme

This paper proposes a new bit-map-based file storage mapping mechanism for multi-cloud and multi-replica settings.

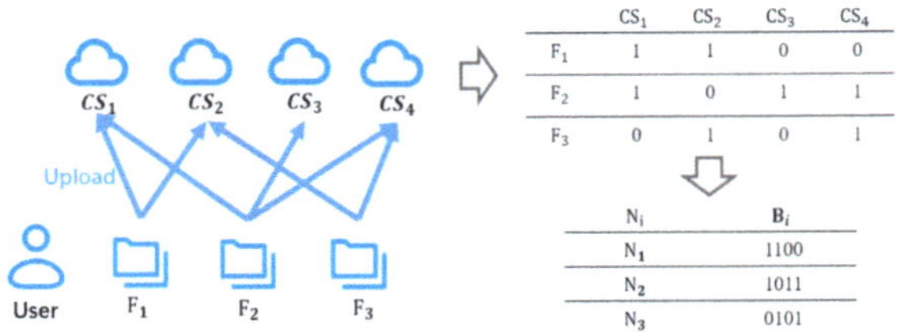

Fig. 2. Schematic diagram of file storage mapping structure

Assume there are m cloud servers, each assigned a unique integer index. The user uploads n files to the cloud combiner. Each file has a filename and a bit-map of length $m/8$ bytes. Specifically, if the i-th bit is *1*, the file should be stored on the cloud server with index i; otherwise, it is not. The user constructs a table T consisting of tuples $T_i = (N_i, B_i)$, where N_i is the hash value of the file name, and B_i is the binary data of the bit-map. This table is then transmitted to the cloud combiner. The combiner stores the files on the cloud servers based on the table T. For example, assume that three files are to be stored on four cloud servers. If the bit-map of file F_1 is 1100, it means F_1 is stored on CS_1 and CS_2, as shown in Fig. 2. This structure reduces storage space usage and allows for quick file location through bitwise operations. It is also scalable.

The following is a detailed description of the specific auditing algorithm.

Setup: The algorithm is run by the KGC.

(1) Select a bilinear map e: $G_1 \times G_1 \to G_2$,where G_1 and G_2 are two multiplicative groups of order q. Let g be the generators of G_1. The KGC randomly selects a number $x \in Z_q^*$ as its private key and computes $Y = g^x$.
(2) The KGC selects a cryptographic hash function.$h : \{0.1\}^* \to G_1$, and defines a pseudorandom function $f : Z_q^* \times \{1,2, \ldots, n\} \to Z_q^*$ and a pseudorandom permutation $\pi : Z_q^* \times \{1,2, \ldots, n\} \to \{1,2, \ldots, n\}$.
(3) The KGC publishes the system parameters $pp = \{q, g, h, G_1, G_2, Y, f, \pi\}$.

Key Generation: The algorithm is executed by the KGC.

(1) The user provides their identity ID to the KGC. The KGC randomly selects r $\in Z_q^*$ and computes $R = g^r, k = r + xh(ID, R) mod q$.
(2) The KGC sends the private key $k_{ID} = (R, k)$ to the user through a secure channel.
(3) The user can verify the correctness of the received private key by checking if the following equation holds: $g^k = R \cdot Y^{h(ID,R)}$.

Tag Generation: The algorithm is executed by the user. Assume there are m cloud servers, denoted as $CS = (CS_1, CS_2, \ldots, CS_m)$.

(1) The user uploads n files $F = (F_1, F_2 \ldots, F_n)$ to multiple cloud servers. And the user calculates the tuple $T_i = (N_i, B_i)$ as mentioned at the outset of Sect. 5, and constructs table T. Subsequently, the user locally saves the table T and sends it to the cloud combiner. The combiner stores the files according to the table T and saves the table T.
(2) For each file F_i where $1 \leq i \leq n$, the user splits the file into s parts, $m_i = \{m_{i1}, m_{i2}, \ldots, m_{is}\}$. For each chunk F_i, user generates a tag $\sigma_{ij} = \left[h(N_i, B_i, j) \cdot g^{m_{ij}}\right]^k$.
(3) Upload the files $F = (F_1, F_2 \ldots, F_n)$ and set of tags $\{\sigma_{ij}\}_{1 \leq i \leq n,\ 1 \leq j \leq s}$ to the cloud.

Challenge Generation: The algorithm is executed by the TPA.

(1) The user sends the challenge file name $N^* = h(name^*)$ and R to the TPA.
(2) The TPA selects a value c, where $1 \leq c \leq s$, as the number of challenged data blocks. The TPA randomly selects two keys $k_1, k_2 \in Z_q^*$. Here, k_1 is the key for the pseudorandom permutation π, and k_2 is the key for the pseudorandom function f.
(3) The TPA sends the challenge $chal = \{N^*, c, k_1, k_2\}$ to the cloud combiner.

Proof Generation: The algorithm is executed by the cloud combiner and the cloud servers.

(1) The cloud combiner locates the index i of the file N^* and the cloud server CS_x based on the tuple $T_i = (N_i, B_i)$. Then, the combiner calculates $\alpha_l = \pi_{k1}(l)$, $\beta_l = f_{k2}(l)$ and $h_l = h(N_i, B_i, \alpha_l)$, and sends the challenge $chal = \{N_i, c, \alpha_l, \beta_l\}$ to CS_x.
(2) The corresponding cloud server CS_x calculates the proof: $\delta_x = \prod_{l=0}^{c} \sigma_{i\alpha_l}{}^{\beta_l}$ and $\mu_x = \sum_{l=0}^{c} \beta_l \cdot m_{i\alpha_l}$.
(3) The cloud returns the proof set $proof_x = \{\delta_x, \mu_x, h_l\}$ to the TPA.

Proof Verification: The TPA creates an m-bit string A to record the audit results and verifies whether the proofs returned by the cloud are correct.

$$e(\delta_x, g) = e\left(g^{\mu_x} \cdot \prod_{l=1}^{c} h_l{}^{\beta_l}, R \cdot Y^{h(ID,R)}\right)$$

If the equation of $proof_x$ holds, the TPA writes a 1 in the x-th position of A, indicating that file is completely stored; otherwise, the TPA writes a 0, indicating that the file has been tampered with or corrupted. Positions in A where no proof is received are filled with 0. Subsequently, the TPA sends $\{N^*, A\}$ to the user, who can compare A with B_i to determine which cloud servers have completely stored the file.

6 Security Analysis

6.1 Correctness

Theorem 1: If the user, the TPA, the cloud servers and the cloud combiner properly carry out the protocol, and the data integrity proof formula holds, then the scheme is considered to meet the requirements for audit correctness.

Based on the bilinear pairing, the process of data integrity proof is as follows:

$$e(\delta_x, g) = e\left(\prod\nolimits_{l=0}^{c} \sigma_{i\alpha_l}{}^{\beta_l}, g\right)$$

$$= e\left(\prod_{l=0}^{c} \left[h(N_i, B_i, \alpha_l) \cdot g^{m_{i\alpha_l}}\right]^{k \cdot \beta_l}, g\right)$$

$$= e\left(g^{\sum_{l=0}^{c} \beta_l \cdot m_{i\alpha_l}} \cdot \prod_{l=1}^{c} h_l{}^{\beta_l}, g\right)^k$$

$$= e\left(g^{\mu_x} \cdot \prod_{l=1}^{c} h_l{}^{\beta_l}, R \cdot Y^{h(ID,R)}\right)$$

6.2 Privacy Protection

Theorem 2: The TPA is an honest but curious entity. During the auditing process, the TPA is unable to obtain the file information.

Proof: Suppose adversary A is synonymous with the TPA and is curious about file. It can be proven that the TPA cannot obtain specific information.

(1) A receives the public parameters sent by challenger C. Due to the DL problem, A cannot crack the key x based on the public parameters $Y = g^x$ and the generator g.
(2) A receives the encrypted filename $N^* = h(name^*)$ sent by the challenger C. Because the hash function is unidirectional, A cannot reverse deduce the real filename.

In summary, a curious TPA cannot access any private details or file information during the auditing process. Therefore, the proposed scheme ensures data privacy.

7 Performance Analysis

In this section, we evaluate the performance of our proposed scheme and compare its experimental results with those of the schemes proposed by Wang H. et al. [13] and Wang M. et al. [9].

In the performance analysis, we defined following notations: H for hash computation, Mul for multiplication, Pow for exponentiation, Add for addition, and $Pair$ for bilinear pairing. Assuming there is one file with n data blocks, and c challenge blocks. The computational costs of the aforementioned schemes were compared through theoretical analysis, as presented in Table 1.

Table 1. Computational Cost Comparison

Phase	Our scheme	Wang H. et al. [13]	Wang M. et al. [9]
Tag Generation	$(n+1) \cdot H + 2n \cdot Pow$	$(n+1) \cdot Pow + n \cdot (Mul + 2H)$	$(2n+1) \cdot (H + Pow) + 2n \cdot Mul$
Proof Generation	$c \cdot (Pow + Mul) + H$	$c \cdot (2Mul + Pow + Add)$	$c \cdot Pow + (3c+1) \cdot Mul + (c-1) \cdot Add$
Proof Verification	$2Pair + (c+1) \cdot Pow + c \cdot Mul$	$2Pair + (c+n+1) \cdot Pow + (c+n) \cdot Mul$	$2Pair + (c+1) \cdot M + H + (c+2) \cdot Pow$

The simulation for this experiment is run on the Ubuntu 22.04.5 system, configured with an Intel Core i7-12650H CPU 2.68 GHz, and 3.8 GB of RAM. The experiment utilizes the GMP and PBC libraries and is implemented in C language. The experiment is conducted in 3 phases: Tag generation, proof generation, and proof verification. The specific experimental results are shown in Fig. 3.

The experimental results show that compared with the schemes of Wang H. et al. [13] and Wang M. et al. [9], our scheme has lower time costs in the three stages of tag generation, proof generation, and proof verification. The efficiency advantage is particularly evident when dealing with a large number of challenge data blocks.

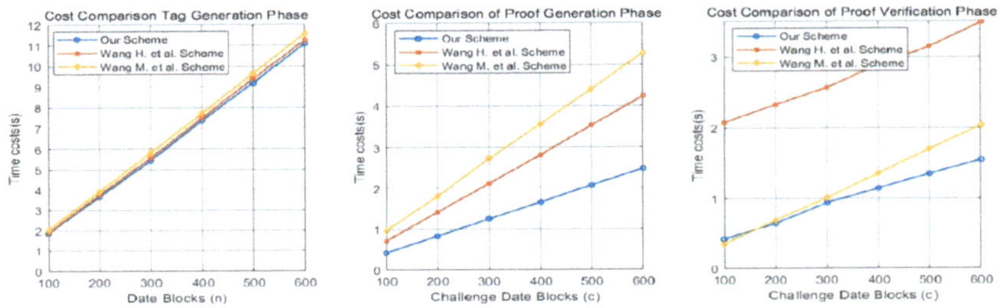

Fig. 3. Comparison of Costs in Three Phases

8 Conclusion

This paper proposes a multi-cloud multi-replica storage auditing scheme based on a cloud combiner, aiming to ensure data integrity and security. Theoretical analysis and experimental results demonstrate that the scheme can effectively and reliably conduct data auditing across multi-cloud and multi-replica environments.

Acknowledgments. The work is supported by the National Key R&D Program of China (No.2023YFB2703700), the National Natural Science Foundation of China (Nos. U21A20465, 62302457, 62172292), the Zhejiang Provincial Natural Science Foundation of China (No. LQ24F020008), the Program for Leading Innovative Research Team of Zhejiang Province (No. 2023R01001), the Fundamental Research Funds of Zhejiang Sci-Tech University (No. 22222266-Y) and the "Pioneer" and "Leading Goose" R&D Program of Zhejiang (No. 2023C01119).

References

1. Marston, S., Li, Z., Bandyopadhyay, S., Zhang, J., Ghalsasi, A.: Cloud computing - the business perspective. Decis. Support. Syst. **51**(1), 176–189 (2011)
2. Shen, J., Shen, J., Chen, X., Huang, X., Susilo, W.: An efficient public auditing protocol with novel dynamic structure for cloud data. IEEE Trans. Inf. Forensics Secur. **12**(10), 2402–2415 (2017)
3. Wang, H., He, D., Fu, A., Li, Q., Wang, Q.: Provable data possession with outsourced data transfer. IEEE Trans. Serv. Comput. **14**(6), 1929–1939 (2019)
4. Ateniese, G., et al.: Provable data possession at untrusted stores. In: Proceedings of 14th ACM Conference on Computer and Communications Security, pp. 598–609. ACM, New York (2007)
5. Juels, A., Kaliski, B.S.: PORs: proofs of retrievability for large files. In: Proceedings of 14th ACM Conference on Computer and Communications Security, pp. 584–597. ACM, New York (2007)
6. Wang, C., Wang, Q., Ren, K., Lou, W.: Privacy-preserving public auditing for data storage security in cloud computing. In: Proceedings of IEEE INFOCOM, pp. 1–9. IEEE, San Diego (2010)
7. Li, J., Zhang, L., Liu, J.K., Qian, H., Dong, Z.: Privacy-preserving public auditing protocol for low-performance end devices in cloud. IEEE Trans. Inf. Forensics Secur. **11**(11), 2572–2583 (2016)

8. Oqaily, M., et al.: SegGuard: segmentation-based anonymization of network data in clouds for privacy-preserving security auditing. IEEE Trans. Dependable Secur. Comput. **18**(5), 2486–2505 (2021)
9. Wang, M., Yu, J., Shen, W., Hao, R.: Privacy-preserving time-based auditing for secure cloud storage. IEEE Trans. Inf. Forensics Secur. **19**, 7866–7878 (2024)
10. Curtmola, R., Khan, O., Burns, R., Ateniese, G.: MR-PDP: multiple-replica provable data possession. In: Proceedings of 28th International Conference on Distributed Computing Systems, pp. 411–420. IEEE, Beijing (2008)
11. Barsoum, A.F., Hasan, M.A.: On verifying dynamic multiple data copies over cloud servers. Cryptol. ePrint Arch., Report 2011/447 (2011)
12. Tian, H., Wang, M., Quan, H., Chang, C.C., Vasilakos, A.V.: TEEMRDA: leveraging trusted execution environments for multi-replica data auditing in cloud storage. Comput. Secur. **150**, 104250 (2025). ISSN 0167-4048
13. Wang, H.: Identity-based distributed provable data possession in multicloud storage. IEEE Trans. Serv. Comput. **8**(2), 328–340 (2015)
14. Lan, C., Li, H., Wang, C.: Analysis of the comments on "Identity-Based Distributed Provable Data Possession in Multicloud Storage." IEEE Trans. Serv. Comput. **14**(1), 44–46 (2021)

DeepWaste: Deep Learning-Based Waste Classification

Duc-Kien Bui[1], Tien-Thanh Do[1](✉), Ngoc-Thanh Pham[1], Ngoc-Anh Le[2], Kim-Thai Dinh[1], Phuong-Thao Le[1], Phuong-Thao Nguyen[1], Minh-Quan Vu[1], and Phuong Anh Nguyen[2]

[1] International School Viet Nam National University, Ha Noi Thanh Xuan, Hanoi 124000, Viet Nam
thanhdo@vnuis.edu.vn

[2] Swinburne University of Technology (Vietnam), 2 Duong Khue Street, Mai Dich Ward, Hanoi 10000, Viet Nam

Abstract. This paper is based on the research process of applying the YOLO model to identify and classify waste. In the self-collected dataset, we tested 2 versions: YOLOv11 and YOLOv12. The test results show that two models have a reasonably high average accuracy of about 71–72% mAP50-95 to identify and classify waste types in the test dataset, a low error rate, and stable performance. Through the research, the reader can see the great uses of artificial intelligence in general and the YOLO model in particular in the field of waste identification and classification. Therefore, this research can become a reference for future research on the application of computer vision in image identification and classification.

Keywords: Classification · YOLO · object detection · deep learning · computer vision

1 Introduction

Vietnam is facing serious waste and pollution problems. Every year, over 3.1 million tons of trash are released into the environment. Only 20% of this waste is properly treated, while the rest is either buried or burned, causing harmful effects on air, water, and soil quality [1]. To solve this issue, trash classification is important to support recycling and reduce environmental damage.

However, in many places, people still mix recyclable and non-recyclable waste. This creates confusion and lowers the efficiency of waste treatment. In recent years, computer vision and artificial intelligence (AI) have been used to help classify waste automatically. In recent years, computer vision and artificial intelligence (AI) have been used to help classify waste automatically. In the current model, YOLO (You Only Look Once) is one of the frequency algorithms to detect objects. YOLO can detect and classify objects in real time with only one scan of the image. The accuracy of the model has improved strongly, from 18–21% in YOLOv1 to 56.8% in YOLOv7 [2].

J. S. Park et al. (Eds.): FutureTech 2025, LNEE 1483, pp. 58–68, 2025.
https://doi.org/10.1007/978-981-95-1999-6_8

In early 2025, a new version called YOLOv12 was released with better performance in object detection and image understanding. Unlike earlier versions, YOLOv12 includes improvements such as faster attention and better training strategies. However, very few studies have tested YOLOv12 for real-world problems like waste classification, especially in complex environments.

This study focuses on evaluating the effectiveness of YOLOv11 and YOLOv12 for waste classification using images collected from different sources. The system detects six diverse labels of waste: cardboard, e-waste, glass, metal, paper, and plastic. The experiment uses a dataset of 8,499 labeled images with diverse angles and lighting conditions. After training and testing the models, the results are compared based on accuracy, precision, recall, and speed.

The main contributions of this study are:

- A total of 8,499 annotated images were compiled into a dataset by merging information from three publicly available sources. This dataset includes six diverse labels: cardboard, electronic waste, glass, metal, paper, and plastic. We carefully filtered and preprocessed the selected images to ensure both high quality and class balance. Two versions of the YOLO model —YOLOv11 and YOLOv12—were tested on the same dataset to compare their performance in detecting and classifying waste.
- A detailed comparison between YOLOv11 and YOLOv12 was made using precision, recall, mAP50, and mAP50-95. YOLOv12 showed better results in most categories, especially in locating objects more accurately.
- The model was trained using Google Colab with a GPU L4. The paper also explores how the model can work on low-power devices and smart systems in real-world use.

2 Material and Method

2.1 Dataset

In this paper, we used data that we collected and processed images to build a dataset of 8,499 labeled images, including categories: 1130 cardboards, metal, 1796 glass, 2897 e-wastes, 1091 papers, and 1988 plastics. This dataset is synthesized from three different sources, namely the 'E-Waste Image Dataset' by Akshat Tamrakar, the 'Garbage Dataset' by Suman Kunwar, and the 'Trash net' by Garythung. After collection, we carefully filter out blurred, poor-quality images and only retain clear images to ensure high accuracy during the labeling process. The post-processed data set is divided into three parts: 70% for training, 15% for validation, and 15% for testing, which helps to ensure that the model has enough data to learn and generalize well on unseen data. More information on the dataset is provided in Table 1.

Table 1. Dataset distribution

Category	Training	Validation	Testing
Image	5952	1265	1282

To improve the model's diversity and generalization, we applied various image processing techniques, including horizontal and vertical flipping, rotation ($-45°$ to $+45°$), axial flipping (vertical and 180°), Gaussian noise addition, and contrast adjustment. Additionally, all images were resized to 640×640 pixels and converted to YOLO format, ensuring consistency in annotation with normalized bounding box coordinates.

2.2 Evaluation Metrics

In this study, we use mAP (mAP50, mAP50-90), the number of parameters, and processing speed as the main evaluation criteria.

The number of parameters represents the total number of weights that can be learned in the model, which has an important influence on the feasibility of real-world systems. The mean Average Precision (mAP) is calculated based on two core metrics: precision and recall. True positive rate (TPR), or the proportion of all actual positives that were classified correctly as positives, is also known as recall [3].

$$Recall\ (or\ TPR) = \frac{correctly\ classified\ actual\ positives}{all\ actual\ positives} = \frac{TP}{TP + FN} \tag{1}$$

Precision is the proportion of all the model's positive classifications that are positive. It is mathematically defined as:

$$Precision = \frac{correctly\ classified\ actual\ positives}{everything\ classified\ as\ positive} = \frac{TP}{TP + FP} \tag{2}$$

In there, the indices TP, FP, and FN respectively have the following meanings: True positive (Number of objects correctly recognized), False positive (Number of misidentified objects), and False negative (Number of objects missed). The mAP is calculated by finding the Average Precision (AP) for each class and then averaging over several classes.

$$AP = \int_0^1 Precision(Recall)dRecall \tag{3}$$

$$mAP = \frac{1}{N}\sum\nolimits_{i=1}^{N} APi \tag{4}$$

where N is the number of classes, APi is the average precision of class i. mAP50 denotes the mean Average Precision (AP) when the Intersection over Union (IoU) threshold is set to 0.5. While mAP50-95 extends this index by averaging AP via multiple IoU thresholds, from 0.5 to 0.95 in increments of 0.05. A lower IoU threshold (e.g., 0.5) that allows the model to accept the bounding box that overlaps with a real object, so the model becomes more tolerant of minor localization errors.

3 Yolo Model

3.1 CNN

Convolutional Neural Network (CNN) is one of the deep learning models. The characteristic of the CNN network is to use many convolutional layers to find image features without humans specifying each pixel. Simply put, the computer is the human, and convolution acts as the eyes to distinguish the difference from what is seen. The CNN model architecture usually consists of layers: input layer, convolutional layer, activation layer, pooling layer, connected layer, flattening output layer [4]. Each input picture will go through several convolutions and kernels. Each kernel looks for a different feature in the image and generates a separate feature map. After each convolutional layer, we apply an activation function to introduce non-linearity into the network, helping the network learn complex features. The Fully Connected Layer is responsible for taking input from previous layers and producing a final output, such as classification or prediction. Figure 1 below illustrates how a CNN works to classify pictures.

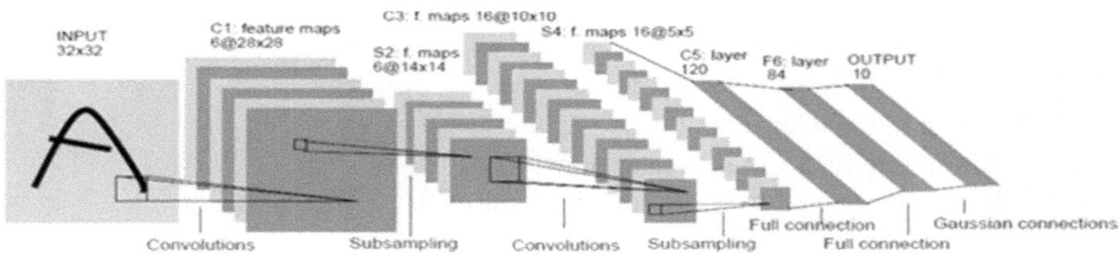

Fig. 1. The architecture of CNNs

CNN is a multi-layer Convolution network combined with non-linear activation functions such as ReLU, and tanh to activate the weights in the nodes [5]. While the Feedforward Neural Network (FNN) model transmits data directly from input to output, that is, each input neuron (input node) to each output neuron in the next layers, the layers of the CNN model will be connected through the convolution mechanism. The characteristic of the CNN network is to use a convolutional layer to search for image features, so convolution is the backbone of the CNN network.

Unlike ANN, CNN uses small matrices to scan through each region of the image to transform the image to obtain features. The convolution layer works based on convolution operations. Usually, in digital signal processing problems, a continuous convolution formula is used, but the input data of CNN is a numeric matrix, so we will apply the discrete convolution formula (5):

$$(F * G)(i, j) = \sum_{m} \sum_{n} F(m, n)G(i - m, i - n) \tag{5}$$

The operation of the discrete convolution process is similar to continuous convolution, the difference is that we work with numerical matrices instead of continuous functions. The formula to process the image matrix in the discrete function is F(m,n), where pixels are collected by reversing the kernels G(i-m, j-n) and scanning via each position of The biggest difference in the discrete convolution formula is the that after multiplying each corresponding element, the sum is added.

This process is repeated until all the small areas of the image are scanned. One convolution layer can use a filter to create several feature maps. Each kernel searches the different aspects in the picture by sliding the kernel over the input picture, multiplies each element by the respective value in each scanning position, and summarizes together and creates a separate feature map [6].

If we only use simple operations such as matrix multiplication or addition, CNN is just a simple linear model that cannot handle complex relationships in the data. Because of that, after each convolution layer, we apply an activation function to send the nonlinearity into the neuron, so the model can identify and process the aspect of the picture, such as the shape, color, or texture. This action is to guarantee that the neural network can learn the diverse features of the input data. The most common in CNN networks is the ReLU activation function (6) because it is suitable for most hidden layers in the network, not specifically for binary classification problems like the Sigmoid function.

$$f(x) = \max(0, x) \tag{6}$$

where

- x is the input to the neuron.
- The function returns x if x is greater than 0.
- If x is less than or equal to 0, the function returns 0.

The formula can also be written as:

$$f(x) = \begin{cases} x, & if\ x > 0 \\ 0, & if\ x \leq 0 \end{cases} \tag{7}$$

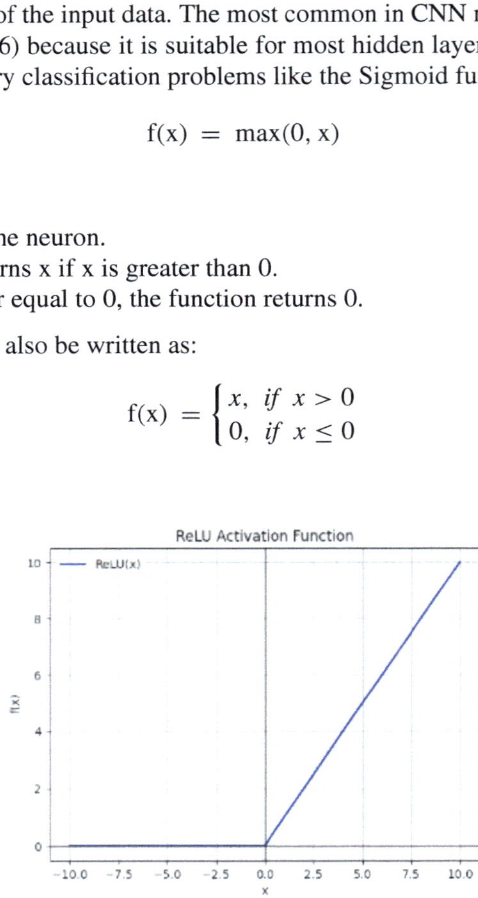

Fig. 2. ReLU Function [7]

With other activation functions, when x has a non-fixed value that can be too large or too small, the derivative is close to 0, causing the network to learn very slowly. However, for the ReLU function, the derivative is always 1 if $x > 0$, so the gradient is not lost when propagating back (Fig. 2).

It is not only used for operation on the input data after the convolution layer, but in the classification task, the activation function is used at the last step, so the network can predict the proportion of each class.

The output of the network is a vector of real numbers that goes through the convolution layer and the pooling layer can receive different values. To ensure the sum proportion of all layers is equal to 1, the Softmax function converts this value into a probability range from 0 to 1 [8].

For example, in an image classification problem with three classes (cat, dog, bird), after applying Softmax, the network might output:

$$\text{Cat} = 0.7, \quad \text{Dog} = 0.2, \quad \text{Bird} = 0.1$$

"Cat" has the highest proportion, so the neuron network will classify it as "cat". The pooling is used to decrease the number of neurons (max_pooling and average_pooling) to simplify the output. While max_pooling chooses the highest value in the feature map, average_pooling calculates the average value of that feature map [9]. The fully connected layer (8) receives the input data from the previous layer to generate the final result, such as classification and prediction.

$$y = f\left(\sum w_i x_i + b\right) \tag{8}$$

After the transmitted image is processed by the convolutional layer and pooling layer, the image's numerical matrix is flattened into a vector and fed into a fully connected layer, such as a neural network [10].

Next, we need to identify the features extracted by the network's previous convolutional layers from the input image. To classify these features, we employ the softmax function, which necessitates a one-dimensional input. The flattened layer is required for this reason.

3.2 Yolo

YOLOv11 is the version of the YOLO model developed by Ultralytics. YOLOv11 has new improvements to overcome the limitations of previous versions, so it is a great method to solve the problem of object recognition and detection. YOLOv11 achieves much higher accuracy than previous versions when its mAP50-95 reaches about 39%–55%. Even in blurred images or small objects, it can learn accurately because it uses complex loss functions, improved anchor boxes, and advanced data augmentation. In addition, the YOLOv11 model has quite good performance and can process images in a short time of about 2.2ms per image. Cross-Stage Partial (CSP) is an important technique in optimizing YOLO (You Only Look Once) models to improve performance and accuracy in object recognition. YOLOv11 uses the CSP (Cross Stage Partial) architecture combined with PANet (Path Aggregation Network) to improve the detection capabilities of different object sizes and orientations.

3.3 Discussion

As shown in Table 2 and Table 3, YOLOv12 demonstrates improved object detection performance, particularly in terms of higher recall. YOLOv12 has a higher Recall at 0.829 and a higher mAP50-95 at 0.726. In comparison, the YOLOv11 achieved a higher at of 0.893 Precision and mAP50. The YOLOv11 consumes less time to train than YOLOv12, with 1 h and 17 min in my dataset, while YOLOv12 was 1 h and 51 min, but

Table 2. YOLOv11n Result

Class	Images	Instances	P	R	mAP50	mAP50-95
all	1265	1521	0.893	0.822	0.893	0.719
cardboard	172	176	0.936	0.916	0.96	0.844
e-waste	404	446	0.882	0.906	0.93	0.75
Glass	208	295	0.901	0.776	0.871	0.65
metal	117	147	0.838	0.687	0.794	0.612
paper	129	143	0.902	0.825	0.891	0.733
plastic	246	314	0.898	0.822	0.911	0.728

Table 3. YOLOv12n Result

Class	Images	Instances	P	R	mAP50	mAP50-95
all	1265	1521	0.886	0.829	0.892	0.726
cardboard	172	176	0.964	0.904	0.963	0.847
e-waste	404	446	0.846	0.906	0.942	0.768
Glass	208	295	0.923	0.776	0.852	0.651
metal	117	147	0.827	0.707	0.787	0.609
paper	129	143	0.887	0.846	0.895	0.755
plastic	246	314	0.868	0.834	0.913	0.729

YOLOv12 demonstrates better performance when trained on synthetic data. With the higher Recall, YOLOv12 can detect the majority of objects, avoiding missing important subjects even with small trash. The index of mAP50-95 allows the YOLOv12 to detect with high accuracy, decreasing the proportion of detecting other objects as trash.

In this project, YOLOv12 was implemented using Google Colab, an interactive online platform that allows testing various samples and algorithms with reduced processing time. Moreover, GPU L4 was invented to be optimal for deep learning and machine learning; it has enhanced the calculation performance and sped up the training process. Also, GPU L4 uses cloud computing, so strong hardware is unnecessary. This accelerated training not only reduces overall processing time but also ensures that the model achieves optimal performance and high accuracy.

The model is trained with 100 epochs. A high epoch count allows the model to learn deeply, but may cause overfitting. Batch size refers to the number of samples processed before updating the weights. A value of 16 balances training speed and gradient stability. The learning rate and momentum were originally set at 0.01 and 0.937, respectively, but were replaced with 0.001 and 0.9 due to the optimization by AdamW. The weight decay was set to 0.0005 to reduce the magnitude of the weights and prevent overfitting, while the IoU threshold was 0.7, defining the minimum intersection-over-union required for a prediction to be considered accurate.

Table 2 compares the performance of YOLOv12 with the different versions with an index of precision (P), Recall (R), Mean Average Precision at 50% IoU(mAP50), and Mean Average Precision across IoU thresholds (mAP50-95). Due to the small size of the model, YOLOv12n can utilize the GPU provided by Colab, leading to a decrease in processing time when training the model. The model consumes less VRAM, affordable with a system that does not have a strong GPU or run in real-time, such as Security surveillance, driver assistance, or an intelligent alert system (Table 4).

Table 4. Comparison of YOLOv12 variations of the detection

Model	P	R	mAP50	mAP50-95
Yolov12n	0.886	0.829	0.892	0.726
Yolov12s	0.898	0.817	0.887	0.722
Yolov12m	0.901	0.819	0.889	0.723
Yolov12l	0.902	0.12	0.892	0.725
Yolov12x	0.905	0.807	0.897	0.728

YOLOv12n is suitable because it runs better on a small device such as a smartphone or a secure camera. The architecture allows it to process efficiently even on devices that have low performance and still maintain high accuracy.

The model performance is good with the mAP = 0.892 and mAP50-95 = 0.726. The model can successfully detect the major of an object. However, some objects are still missing when the model is detected. The model shows the highest performance when detecting Cardboard and E-waste, but it is difficult to recognize Metal.

The model has high accuracy but is still missing some objects. It operates better at IoU 50% but decreases when the requirement is higher IoU we can see the performance of the model decreases obviously when changing from IoU 50% to IoU 50–95%. E-waste label has a high Recall, but the Precision is lower may lead the confusion with other labels. The glass category has low recall, indicating that the model frequently fails to detect relevant objects, while the metal category shows the weakest performance, with many errors of missing and confusion due to the Recall at 0.707 and the Precision at 0.827. While YOLOv12 shows a good performance in detecting most of the objects, the accuracy in detecting some label—especially Metal—still not optimized. As we can see in Table 3, the Metal label records the lowest Precision and Recall the lowest in all the labels. This proves that the model not only misses some of the metal objects, but also detects Metal label as another trash label frequency, such as E-waste and Glass.

One of the factors contributing to this problem is because of the reflective nature of the metal, which can be diverse due to the light condition, the angle, and the shadow. This may lead to visual noise, which interferes with the aspect extraction process.

Furthermore, the metal object may have a similar shape and texture to other trash. For instance, the crushed cans may resemble shattered glass or small electronic devices. The visual similarity between metal and other materials, along with the lack of variety and representation in training samples for the metal class, poses a challenge for the model in learning clear and distinctive features. Improving this may require more diverse and balanced training data, or the use of attention mechanisms or contrastive learning to better distinguish fine-grained features (Figs. 3, 4 and 5).

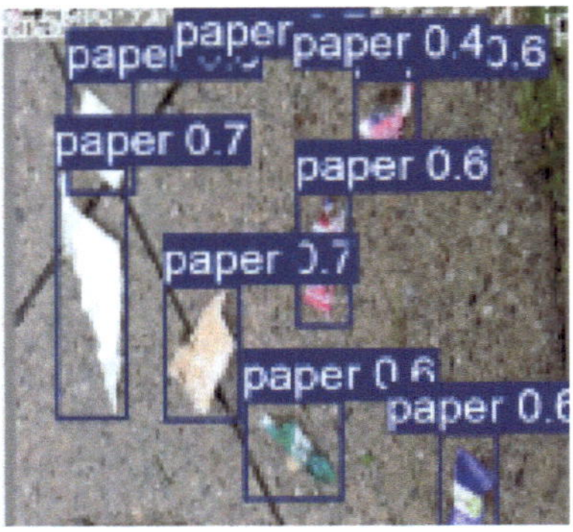

Fig. 3. Result of YOLOv12

Fig. 4. Result of YOLOv12

Fig. 5. Result of YOLOv12

4 Conclusion

In a nutshell, for our purpose is precision, accuracy, and stability, YOLOv12 proves that it is more suitable for this task than YOLOv11. The model especially operates well when detecting objects, but the ability to recognize reflective surfaces and metallic objects still shows some weakness.

The YOLOv12 model operates efficiently, however, it still has some problems when detecting Metal, as we can see via the lower mAP of Metal than other labels. The problem may occur due to the reflection of metal surfaces, directly impacting the model's training process.

Furthermore, the gap between mAP50-95 and mAP50 illustrates that the model works better at lower IoU, but is difficult to process at high IoU. This means that the model can detect the object, but the capacity to recognize the exact position is still not optimized. For future development, to enhance the accuracy as well as the stability of the model, I will focus on optimizing the mAP50-95 so the model can have higher performance in different and complex conditions.

The model can be attached with a high-quality camera to gather images more clearly to detect trash more accurately. Also, we can attach the model to the web so people can detect trash trash to classification and recycling. Moreover, the labels of the dataset need to be balanced to improve the performance of the classification trash. For the extended improvement, the model can be attached to IoT, such as connecting the system with smart trash bins, so it can automatically record and optimize the gathering process.

References

1. Khuc, V.Q., et al.: Exploring strategies to tackle plastic waste pollution: evidence from Vietnamese household survey and Bayesian approach. In: Proceedings of the International Conference on Sustainable Environmental Research (ICSER 2022), pp. 1–4. Springer, Cham (2022)
2. Madde, K., Kumar, N.: Advanced YOLO-based trash classification and recycling assistant for enhanced waste management and sustainability. In: Proceedings of the 2024 International Conference on Smart Waste Management (SWM 2024). LNCS, vol. 13990, pp. 6–7. Springer, Cham (2024)
3. Henderson, P., Ferrari, V.: End-to-end training of object class detectors for mean average precision. In: Lai, SH., Lepetit, V., Nishino, K., Sato, Y. (eds.) Computer Vision – ACCV 2016, ACCV 2016. LNCS, vol. 10115, pp. 198–213. Springer, Cham. https://doi.org/10.1007/978-3-319-54193-8_13
4. Indolia, S., Goswami, A.K., Mishra, P., Asopa, P.: Conceptual understanding of convolutional neural network—a deep learning approach. In: Proceedings of the International Conference on Neural Computation (ICNC 2018), pp. 2–4. Springer, Cham (2018)
5. Nagabushanam, D., Mathew, S., Chowdhary, C.: A study on the deviations in performance of FNNs and CNNs in the realm of grayscale adversarial images. In: Proceedings of the International Conference on Machine Vision (ICMV 2022), pp. 9–12. Springer, Cham (2022)
6. Mukta, M., Jonkman, M.: Using feature maps to unpack the CNN 'Black box' theory with two medical datasets of different modality. In: Proceedings of the International Symposium on Biomedical Computing (ISBC 2023), pp. 3–6. Springer, Cham (2023)

7. Kamalov, F., Nazir, A., Safaraliev, M., Cherukuri, A.K., Zgheib, R.: Comparative analysis of activation functions in neural networks. In: Proceedings of the International Conference on Neural Information Systems (NIS 2021), pp. 3–6. Springer, Cham (2021)
8. Iwana, B., Kuroki, R., Uchida, S.: Explaining convolutional neural networks using softmax gradient layer-wise relevance propagation. In: Proceedings of the International Conference on Pattern Recognition (ICPR 2019), pp. 4–8. Springer, Cham (2019)
9. Doğan, Y.: Which pooling method is better: Max, Avg, or Concat (Max, Avg). In: Proceedings of the 2023 International Conference on Neural Systems and Applications (ICNSA 2023), pp. 2–4. Springer, Cham (2023)
10. Ghosh, A., Sufian, A., Sultana, F., Chakrabarti, A., De, D.: Fundamental concepts of convolutional neural network. In: Balas, V., Kumar, R., Srivastava, R. (eds.) Recent Trends and Advances in Artificial Intelligence and Internet of Things. ISRL, vol. 172, pp. 519–567. Springer, Cham (2020). https://doi.org/10.1007/978-3-030-32644-9_36

Federated Unlearning: Efficient Data Removal Strategies and Challenges in Privacy-Preserving Machine Learning

Qingyu Tan[1], Yan Li[1], Jungho Kang[2], and Byeong-Seok Shin[1(✉)]

[1] Inha University, 100 Inha-ro, Michuhol-gu, Incheon, South Korea
bsshin@inha.ac.kr

[2] Baehwa Women's University, 52 Baehwa-ro, Jung-gu, Seoul, South Korea

Abstract. In recent years, the Right to Be Forgotten (RTBF) has emerged as a critical issue in promoting digital trust and AI safety, highlighting the necessity of effectively removing identifiable information from trained machine learning models. This necessity has stimulated the development of Machine Unlearning (MU), enabling ML models to selectively remove specific data contributions without retraining from scratch. Building upon MU, Federated Unlearning (FU) has been developed to address data deletion challenges inherent in Federated Learning (FL) scenarios, granting federated models the capability to selectively remove the contributions of specific clients or identifiable client-related data. However, the unique characteristics of federated learning pose additional challenges that cannot be fully addressed by existing centralized unlearning approaches. In this survey, we first provide a concise overview of the FU framework, followed by a detailed introduction to the concepts and definitions related to MU and FU. We then summarize the key challenges currently faced in federated unlearning research. Furthermore, we systematically classify existing federated unlearning algorithms according to their specific tuning objectives. Finally, we outline several promising research directions for future investigations in the field of federated unlearning.

Keywords: Federated unlearning · digital trust · AI safety

1 Introduction

In recent years, data privacy regulations such as the GDPR [1] and CCPA [2] have introduced stringent requirements for data handling, including the right to be forgotten (RTBF). RTBF grants individuals the right to request the removal of their personal data from service providers, ensuring that their information is neither stored nor utilized in any form. While this principle is well-defined for centralized data storage systems, its implementation in machine learning (ML) models remains challenging, as models inherently retain information from training data. Consequently, Machine Unlearning (MU) [3] has emerged as a crucial research direction, focusing on selectively eliminating the influence of specific data points from a trained model without necessitating full retraining.

B.-S. Shin—This work is supported by Inha University.

J. S. Park et al. (Eds.): FutureTech 2025, LNEE 1483, pp. 69–74, 2025.
https://doi.org/10.1007/978-981-95-1999-6_9

MU techniques are primarily designed for centralized ML settings, where data is directly accessible for modification or removal. However, with the emergence of Federated Learning (FL) [5]—a decentralized paradigm that facilitates collaborative model training across multiple devices without exchanging raw data—conventional unlearning approaches become impractical. The distributed, privacy-preserving nature of FL introduces significant challenges, as data remains localized on user devices, limiting direct intervention for data removal. To address these challenges, Federated Unlearning (FU) [4] has emerged as a novel framework designed to efficiently remove an FL client's contribution or any identifiable information associated with their partial data from a federated model, while simultaneously preserving system privacy, computational efficiency, and model integrity.

Despite its importance, FU remains an underexplored area with critical open questions regarding computational efficiency, security guarantees, and unlearning effectiveness. In this paper, we provide a comprehensive review of FU techniques, categorize existing methods, and analyze their advantages and limitations. Furthermore, we discuss emerging solutions that seek to balance privacy, efficiency, and accuracy in federated settings, offering insights into future research directions.

2 Preliminaries and Backgrounds

2.1 Machine Unlearning

The goal of ML is to extract valuable knowledge through the analysis and modeling of extensive data. In contrast, MU [6–10] endows models with the capability to actively "forget" specific data points or subsets. Specifically, MU techniques aim to adjust trained model parameters in such a way that the resulting model closely approximates the state in which the specific data had never been included in the training set, thereby avoiding the substantial computational overhead associated with complete model retraining. Consequently, machine unlearning provides ML models with the ability to actively "forget" specific data, thereby satisfying data privacy requirements, enhancing model security, and ensuring regulatory compliance.

The difficulty of MU varies depending on the underlying ML scenarios. In settings where all data is centrally available, machine unlearning can effectively leverage accessible data and model parameters to facilitate unlearning. For example, numerous unlearning methods have been proposed in centralized learning environments, tailored specifically to different models and application requirements. These methods include targeted unlearning approaches for regression models [6], rapid unlearning mechanisms for recommender systems [7], and few-shot unlearning techniques designed for scenarios with limited data availability [8].

However, with increasingly decentralized data storage and stricter privacy constraints, traditional centralized unlearning approaches are becoming inadequate to meet contemporary needs. Particularly, in scenarios involving fully decentralized data, where clients cannot directly communicate or share raw data, conventional unlearning methods [6–8] face significant limitations. These practical challenges have driven the emergence of a new research direction known as federated unlearning [4], aiming to efficiently remove a client's contribution from federated models while adhering to strict privacy and communication constraints.

2.2 Federated Unlearning

FU operates by updating federated models to selectively erase the impact of specific clients or subsets of client data, addressing privacy concerns inherent in decentralized learning scenarios [9]. Generally, the FU workflow involves two main steps. Initially, a global model is trained collaboratively by multiple clients without sharing raw data. Formally, given a set of clients $U = \{u_1, u_2, \cdots, u_N\}$, FL produces a global model ω trained on the distributed datasets $\{D_1, D_2, \cdots D_N\}$, where each D_i represents the local dataset of client u_i.

Subsequently, upon receiving a request from client u_i to remove its data contribution $D_i' \subseteq D_i$, the central server initiates the unlearning process. Federated unlearning seeks a modified global model ω' that satisfies:

$$\left|T\left(\omega'\right) - T_{u\setminus u_i}\left(\omega^*\right)\right| \le \epsilon, \quad (1)$$

where $T(\cdot)$ denotes the data distribution of the updated model, $T_{u\setminus u_i}(\omega^*)$ represents the target model distribution achieved by retraining from scratch on the datasets of all remaining clients excluding u_i, and ϵ is a small tolerance threshold quantifying acceptable divergence.

The fundamental objective of FU is thus to achieve efficient and precise data erasure from federated models, ensuring that the updated model closely approximates the scenario where the removed client's data was never involved in training, while simultaneously adhering to strict privacy and communication overhead constraints.

2.3 Challenges of Federated Unlearning

While federated unlearning offers a promising solution for enabling data removal in distributed environments, its practical implementation presents several significant challenges [10–12]. One of the primary difficulties stems from the inherent data heterogeneity across federated clients, which complicates the unlearning process. Each client maintains a distinct local data distribution, and the global model continuously aggregates contributions from multiple clients over successive training rounds. As a result, eliminating or modifying local updates does not necessarily guarantee the complete removal of a client's data contribution from the global model [10].

Another major challenge arises from the iterative and cumulative nature of federated training, which complicates the identification and removal of specific data influences [11]. Since parameter updates from multiple clients are aggregated over time, dependencies between updates become intertwined across different training iterations. In the absence of an effective mechanism to track and manage historical training information, unlearning may induce unintended consequences, such as catastrophic forgetting [12], leading to substantial degradation in model performance and generalization capability.

Additionally, FU must address fundamental constraints related to communication bandwidth, computational efficiency, and privacy preservation in FL systems [13]. The removal of client contributions must be executed with high precision while minimizing adverse effects, such as increased communication overhead, weakened privacy guarantees, or deterioration in model accuracy.

Overall, FU aims to precisely and efficiently eliminate a client's data contribution from a federated model while preserving model integrity, communication efficiency, and privacy protection. The distinct challenges associated with FU establish it as a crucial and rapidly evolving research domain in privacy-preserving machine learning, highlighting the need for further advancements in unlearning methodologies to enhance their robustness, scalability, and applicability [14].

3 Federated Unlearning Methods

In this section, we provide a comprehensive overview of current FU algorithms by categorizing them according to their specific modification targets during the unlearning process. Broadly, existing FU algorithms can be divided into two categories: global-model-oriented unlearning [15–18] and local-model-oriented unlearning [19–22].

Global-model-oriented unlearning algorithms directly modify the parameters of the global federated model to remove the contributions from specific clients or datasets. Such approaches typically involve centralized parameter adjustments executed at the server side. The central server recalibrates or perturbs the aggregated global parameters to approximate a scenario in which the specific client's data never participated in training, thus achieving effective unlearning without explicitly requiring client-side interactions.

In contrast, local-model-oriented unlearning algorithms focus on adjustments at the individual client's local model updates rather than directly modifying the global model. Due to privacy constraints in federated learning, each client u_i maintains its local model parameters ω_i trained independently on their local dataset D_i. Thus, the goal of local-model-oriented unlearning is to revise the local model updates ω_i of the client whose data is to be forgotten. This procedure can be represented as:

$$\omega_i' = \mathcal{U}\left(\omega_i, D_i^{forget}\right), \tag{2}$$

where $\mathcal{U}$ denotes the unlearning operation applied locally on client i's model parameters ω_i, and D_i^{forget} represents the subset of the client's local data targeted for removal. The adjusted local updates ω_i' are subsequently aggregated by the central server, resulting in a modified global model ω'. This local-oriented approach minimizes disruption to the global model and reduces communication overhead during unlearning.

Both categories of federated unlearning algorithms face common challenges, such as ensuring precise removal of client contributions, avoiding performance degradation or catastrophic forgetting, and maintaining compliance with stringent privacy and communication constraints.

4 Discussions and Promising Directions

While initial studies have explored data removal in federated learning, FU remains in its early stages with several promising research directions. One key area is improving data compliance and secure circulation in regulated data marketplaces, enabling data owners to manage and withdraw contributions efficiently. Additionally, cross-domain FU must address data heterogeneity and inter-domain communication constraints to ensure accurate removal of diverse data influences [23].

Another important direction is personalized FU [24], which accommodates fine-grained unlearning requests, such as removing specific user preferences instead of entire datasets. Moreover, FU must be adapted for specialized applications in resource-constrained environments like healthcare, finance, and edge computing, where privacy and efficiency are critical [25].

Future advancements in FU should focus on enhancing compliance, refining unlearning techniques for diverse environments, and optimizing resource-efficient strategies, ensuring practical deployment across real-world federated learning systems.

5 Conclusions

This survey provides a comprehensive analysis of federated unlearning, examining its objectives, challenges, and methodologies. Key challenges include data privacy, model security, computational efficiency, and communication overhead. By establishing a systematic research framework, this work offers valuable insights for future research in FU. Additionally, we outline promising directions, such as FU in data marketplaces, cross-domain learning, personalized unlearning, and specialized applications. Advancing these areas is crucial for enhancing FU's real-world applicability and ensuring its continued evolution as a vital component of federated learning systems.

References

1. Regulation, P.: General data protection regulation. Intouch **25**, 1–5 (2018)
2. Goldman, E.: An introduction to the California Consumer Privacy Act (CCPA). Santa Clara Univ. Legal Studies Research Paper (2020)
3. Bourtoule, L., et al.: Machine unlearning. In: 2021 IEEE Symposium on Security and Privacy (SP). IEEE (2021)
4. Liu, G., et al.: Federated unlearning. arXiv preprint arXiv:2012.13891 (2020)
5. McMahan, B., et al.: Communication-efficient learning of deep networks from decentralized data. In: Artificial Intelligence and Statistics. PMLR (2017)
6. Tarun, A.K., Chundawat, V.S., Mandal, M., Kankanhalli, M.: Deep regression unlearning. In: International Conference on Machine Learning. PMLR, pp. 33921–33939 (2023)
7. Chen, C., Sun, F., Zhang, M., Ding, B.: Recommendation unlearning. In: Proceedings of the ACM Web Conference, pp. 2768–2777 (2022)
8. Yoon, Y., et al.: Few-shot unlearning by model inversion. arXiv preprint arXiv:2205.15567 (2022)
9. Liu, Y., Xu, L., Yuan, X., et al.: The right to be forgotten in federated learning: an efficient realization with rapid retraining. In: IEEE INFOCOM 2022 - IEEE Conference on Computer Communications, pp. 1749–1758. IEEE (2022)
10. Nguyen, D.C., Ding, M., Pathirana, P.N., et al.: Federated learning for Internet of Things: a comprehensive survey. IEEE Commun. Surv. Tutor. **23**(3), 1622–1658 (2021)
11. Dong, J., Wang, L., Fang, Z., et al.: Federated class-incremental learning. In: Proceedings of the IEEE/CVF Conference on Computer Vision and Pattern Recognition (CVPR), pp. 10164–10173 (2022)

12. Liu, Y., Fan, M., Chen, C., et al.: Backdoor defense with machine unlearning. In: IEEE INFOCOM 2022 - IEEE Conference on Computer Communications, pp. 280–289. IEEE (2022)
13. Liu, Z., Jiang, Y., Shen, J., et al.: A survey on federated unlearning: challenges, methods, and future directions. ACM Comput. Surv. **57**(1), 1–38 (2024)
14. Daluwatta, W., Khalil, I., Edirimannage, S., et al.: UaaS-SFL: unlearning as a service for safeguarding federated learning. IEEE Trans. Netw. Serv. Manag. **22**(2), 1029–1045 (2025)
15. Su, N., Li, B.: Asynchronous federated unlearning. In: IEEE INFOCOM 2023 - IEEE Conference on Computer Communications, pp. 1–10. IEEE (2023)
16. Fraboni, Y., Van Waerebeke, M., Scaman, K., et al.: Sifu: sequential informed federated unlearning for efficient and provable client unlearning in federated optimization. In: International Conference on Artificial Intelligence and Statistics. PMLR, pp. 3457–3465 (2024)
17. Wu, C., Zhu, S., Mitra, P.: Federated unlearning with knowledge distillation. arXiv preprint arXiv:2201.09441 (2022)
18. Zhu, X., Li, G., Hu, W.: Heterogeneous federated knowledge graph embedding learning and unlearning. In: Proceedings of the ACM Web Conference 2023, pp. 2444–2454 (2023)
19. Liu, G., Ma, X., Yang, Y., et al.: FedEraser: enabling efficient client-level data removal from federated learning models. In: 2021 IEEE/ACM 29th International Symposium on Quality of Service (IWQoS), pp. 1–10. IEEE (2021)
20. Gao, X., Ma, X., Wang, J., et al.: Verifi: towards verifiable federated unlearning. IEEE Trans. Dependable Secure Comput. **21**(6), 5720–5736 (2024)
21. Li, G., Shen, L., Sun, Y., et al.: Subspace-based federated unlearning. arXiv preprint arXiv: 2302.12448 (2023)
22. Gong, J., Kang, J., Simeone, O., et al.: Forget-SVGD: particle-based Bayesian federated unlearning. In: 2022 IEEE Data Science and Learning Workshop (DSLW), pp. 1–6. IEEE (2022)
23. Zhang, H., Zhang, Y., Jia, K., et al.: Unsupervised domain adaptation of black-box source models. arXiv preprint arXiv:2101.02839 (2021)
24. Jannach, D., Manzoor, A., Cai, W., et al.: A survey on conversational recommender systems. ACM Comput. Surv. (CSUR) **54**(5), 1–36 (2021)
25. Imteaj, A., Thakker, U., Wang, S., et al.: A survey on federated learning for resource-constrained IoT devices. IEEE Internet Things J. **9**(1), 1–24 (2021)

Wetland Vegetation Identification Model Based on Improved Deeplabv3+ and Contrastive Learning

Xiuhe Yuan[1], Han Li[1], Guoqing Ni[2], Zitong Liu[2], Sheng Miao[2], and Chao Liu[1](✉)

[1] School of Environmental and Municipal Engineering, Qingdao University of Technology, Qingdao 266520, China
{li_han,liuchao}@qut.edu.cn
[2] School of Information and Control Engineering, Qingdao University of Technology, Qingdao 266520, China
smiao@qut.edu.cn

Abstract. Aiming at the problem of insufficient classification accuracy of traditional methods in complex wetland environments, this paper proposes an improved DeeplabV3+ model (Ct-Deeplabv3+) integrating a comparative learning encoder and a dual attention mechanism: a clustering-based comparative learning mechanism is designed at the encoder side to optimize the feature representation in an unsupervised manner, and an improved inflated convolution strategy is combined to expand the sensory field. The decoder side introduces the dual modules of gated context transformation and channel-coordinate attention to strengthen the ability of multi-scale feature fusion and spectral weight assignment. The results of the ablation experiments show that the comparative learning encoder and the dual attention mechanism module can effectively improve the benchmark model accuracy. The Ct-Deeplabv3+ model is applied to the task of semantic segmentation of high-resolution multispectral UAV remote sensing images, and the experimental results show that the accuracy of Ct-Deeplabv3+ on the test set reaches 93.5%, and the recall and Kappa coefficients are 88.4% and 0.9318, which is better than that of the mainstream model. This paper provides a new technical model for high-precision dynamic monitoring of wetland vegetation, which has important application value for ecological restoration and sustainable development.

Keywords: Deep Learning · Unmanned Aerial Vehicle · Contrastive Learning · Attention Mechanisms · Wetland Vegetation

1 Introduction

Wetland vegetation is the ecological cornerstone for maintaining hydrological connectivity, carbon and oxygen balance and environmental purification, and its spatial pattern directly characterizes the health of wetlands [1]. The urbanization process has led to the reduction of its area and fragmentation of its structure, resulting in ecological degradation and loss of function [2]. Accurate monitoring of its spatial pattern and dynamic

J. S. Park et al. (Eds.): FutureTech 2025, LNEE 1483, pp. 75–80, 2025.
https://doi.org/10.1007/978-981-95-1999-6_10

changes is the scientific basis for assessing the ecological status and formulating conservation strategies, and is also a key support for promoting biodiversity and sustainable development.

Early remote sensing monitoring of wetland vegetation mainly relied on visual interpretation or spectral characterization of low-resolution satellite images, which was inefficient and had limited classification accuracy [3]. After the introduction of machine learning techniques, the accuracy of wetland classification has been significantly improved [4]. For example, Wang et al. used SVM algorithm combined with image texture and spectral features to recognize crops and achieved a classification accuracy of 84.26% [5]. In recent years, with the development of high-resolution remote sensing and deep learning, Zhao et al. used the ResNet-18 model to classify Landsat 8 images with an accuracy of 94.36%, demonstrating the effectiveness and accuracy of deep learning methods [6]. In order to cope with the challenges of orientation diversity, domain differences, and detail loss in remote sensing images, researchers have continuously optimized deep learning model architectures. HAssNet proposed by Lv et al. enhances multi-scale feature representations through spatial and channel attention mechanisms, which improves global perception and small target detection [7]; innovative architectures based on the visual Transformer such as Swin Transformer synergizes local and global attention through sliding windows [8]; and the ST-UNet hybrid model fuses the local feature extraction advantage of CNN with the global modeling capability of ViT, which improves the boundary recognition accuracy while controlling the computational complexity [9].

In summary, this paper proposes a deep learning model based on contrast learning encoder to improve the classification accuracy of wetland vegetation. In the process of training the contrast learning encoder, this paper uses a clustering-based approach to design based on the classification categories in constructing the contrast learning positive and negative samples. In addition, inside this deep learning model, a module integrating the dual attention mechanism of GCT and CCA is introduced to increase the detail texture information and reasonably assign spectral weights, and the feature extraction results of the contrast learning encoder and DeeplabV3+ are fused.

2 Methodology

2.1 Data Preparation

In this paper, a Phantom4 multispectral UAV was selected for high-resolution remote sensing data acquisition, which is equipped with a visible lens and five narrowband spectral lenses and RTK positioning module, of which the narrowband spectra include red (R), green (G), blue (B), near-infrared (NIR), and red-edge (RE) light waves. In order to obtain high-quality image data with stable solar altitude angle, the aerial survey time was selected as May-October vegetation period, 10:00–12:00 and 14:00–16:00 every day, and the route planning was completed based on the DJI GS Pro software, and the data acquisition was carried out by using the relative altitude of 100m, and the spatial resolution of the original image was 5cm. In post-processing, the spatial resolution of the remote sensing image was adjusted from 5 cm to 25 cm by down-sampling technique.

2.2 Model Framework

In this paper, we propose Ct-Deeplabv3+,which is optimized based on the DeepLabV3+ architecture (Fig. 1): the encoder side fuses the features of the contrast learning encoder with the high-level semantic features extracted from the Xception65 backbone network, and expands the sensory field by an improved inflationary convolution strategy (1×1 convolution using an inflation factor of 2, and 3×3 convolution using a three-branching inflation factor (12,24,36)) to expand the receptive field for enhanced model context understanding. The decoder side introduces GCT and CCA dual-attention modules to enhance feature focusing, and constructs a U-shaped multilevel fusion path to enhance spatial detail retention by four times 2-fold up-sampling and three times cross-scale feature fusion (integrating the ASPP outputs and the 1/2/1/4/1/8 feature maps of Xception65), combined with the UNet-style hierarchical connectivity mechanism to form a final model that takes into account both global Contextual modeling and high-resolution detail reconstruction fusion architecture.

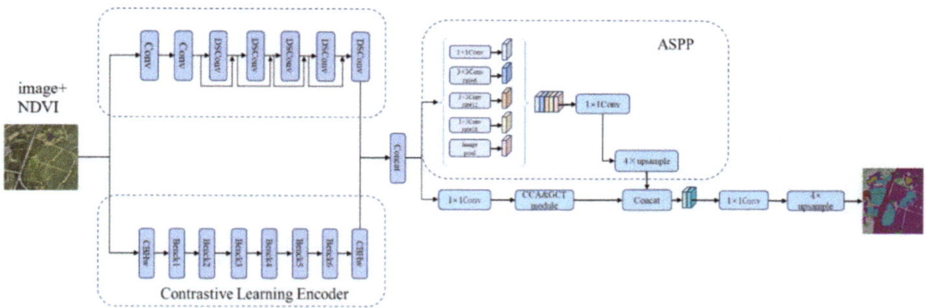

Fig. 1. The framework of proposed model

2.3 Modeling

In this paper, the Ct-Deeplabv3+ model is used for the semantic segmentation task, and the model training is based on the super-resolution reconstructed multispectral remote sensing image dataset, and the classification system is constructed for seven typical feature categories: bare ground, grass, shrubs, trees, aquatic plants, water bodies and buildings. The original image is pre-processed and cut into tif sub-images of $64 \times 64 \times 4$ size, and a total of 10,552 samples are generated, which are divided into training set, validation set and test set in the ratio of 7:2:1. The cross-entropy loss function with Adam optimizer is used in the training process, the initial learning rate is 5×10^{-3}, the minimum learning rate is set to 1% of the initial value, the learning rate momentum update parameter is 0.9, the weight decay coefficient is 1×10^{-4}, and the learning rate is dynamically adjusted using Yolox. After the model is trained for 200 rounds, the segmentation performance is evaluated on the test set for 7 types of features. To evaluate the impact of the contrast learning encoder as well as the GCT and CCA attention modules on the overall model recognition accuracy, we conducted ablation experiments. In addition, in order to verify the performance advantages of the proposed method, we also conducted comparative experiments with mainstream models such as U-Net and SegNet. For model accuracy assessment, Accuracy as well as Recall and Kappa coefficients are used as metrics to evaluate the classification accuracy.

2.4 Result and Analysis

The training loss of Ct-Deeplabv3+ semantic segmentation model is shown in Fig. 2, in the initial training phase of 0–20 rounds, the training loss and validation loss both show a rapid decreasing trend; then from 20 to 120 rounds, the loss decreasing rate gradually slows down; finally after 120 rounds, the training loss and validation loss curves all tend to be stable and maintained at a lower level, and the double loss values tend to converge at the same time, which indicates that the model has completed effective training and reached a stable convergence state.

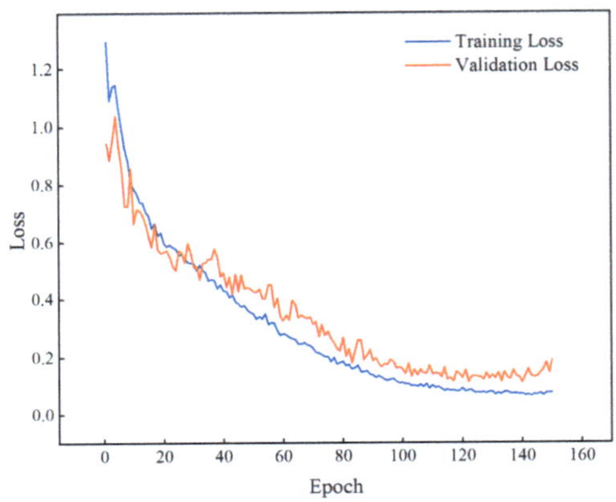

Fig. 2. Modeling loss variation curve

The results of the ablation experiments are shown in Fig. 3a. CDeeplabv3+ A denotes the case of improving the attention mechanism only, and CDeeplabv3+ B denotes the case of improving the contrast learning encoder only. Improving the contrast learning encoder alone improves the recognition accuracy of the ground categories in the dataset, while improving the attention mechanism module alone has a smaller improvement effect. However, using the improved attention mechanism module on top of the improved contrast learning encoder achieves better results.

The results of the comparison experiment are shown in Fig. 3b, Ct-Deeplabv3+ demonstrates optimal performance on the remote sensing dataset, with an Accuracy of 93.5%, a Recall of 88.4%, and a Kappa coefficient of 0.9318, which has a significant accuracy advantage over Deeplabv3+, SegNet, and U-net, and demonstrates better robustness. This demonstrates that the network model proposed in this paper can extract richer feature information from the input image with different convolutional kernel sizes, which compensates for the influence of lightweight feature extraction networks on the segmentation accuracy of semantic segmentation models.

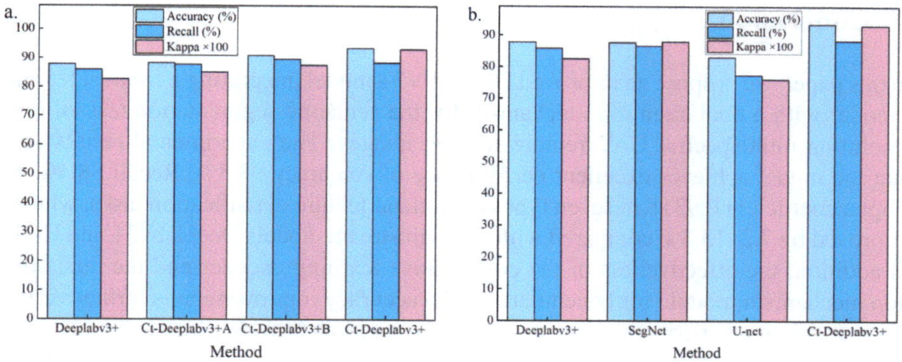

Fig. 3. Model performance in ablation and comparison experiments

2.5 Vegetation Identification

The Ct-Deeplabv3+ model proposed in this paper is used to train semantic segmentation on randomly selected UAV remote sensing tif sub-images. The experimental results show (Fig. 4) that the model not only accurately classifies different feature types, but also exhibits excellent recognition performance for the segmented types of trees, shrubs and grassland wetlands in the vegetation cover area.

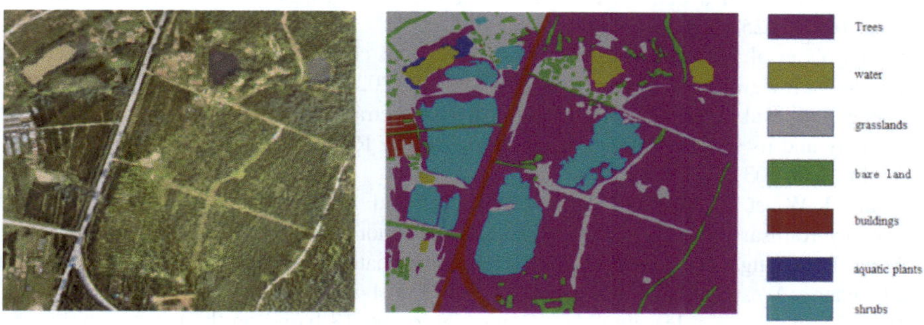

Fig. 4. Vegetation identification results

Semantic segmentation of remote sensing images aims to extract valuable information from complex remote sensing data by combining pixel-level spectral analysis and spatial pattern recognition. The Ct-Deeplabv3+ model not only accurately recognizes spectral features in remote sensing images, but also effectively captures spatial dependencies between the features, thus realizing efficient and accurate recognition of various types of targets. Therefore, the model has a wide range of applications that are not limited to specific types of remote sensing images or geographic areas, which makes Ct-Deeplabv3+ a powerful tool for diverse environmental monitoring tasks, including wetland protection, agricultural assessment, and urban planning.

3 Conclusion

In this paper, we propose an improved DeeplabV3+ model integrating a contrast learning encoder with a dual attention mechanism for the semantic segmentation task of high-resolution multispectral UAV remote sensing images. The experimental results show that the model achieves excellent performance of Accuracy 93.5%, Recall 88.4% and Kappa coefficient 0.9318 in seven types of wetland feature classification tasks, which is improved by 7.2–15.8% compared with the mainstream models Deeplabv3+ and U-Net. In addition, the effectiveness of the comparative learning encoder and the dual attention mechanism module for benchmark model accuracy improvement is further verified through ablation experiments. The results of this research provide a new technical model for high-precision dynamic monitoring of wetland vegetation, and its application can support the early warning of ecological degradation and the formulation of protection strategies. Future research will focus on expanding the model's adaptive ability to multi-temporal and multi-source remote sensing data, and exploring its potential for quantitative assessment of ecological service functions under the context of global climate change.

References

1. Mao, D.H., et al.: National wetland mapping in China: a new product resulting from object-based and hierarchical classification of Landsat 8 OLI images. ISPRS J. Photogramm. Remote. Sens. **164**, 11–25 (2020)
2. Xiong, Y., et al.: Influence of human activities and climate change on wetland landscape pattern—a review. Sci. Total Environ. **879**, 163112 (2023)
3. Cheng, S.Y., et al.: Using ZY1-02D satellite hyperspectral remote sensing to monitor landscape diversity and its spatial scaling change in the Yellow River Estuary. Int. J. Appl. Earth Obs. Geoinf. **128**, 103716 (2024)
4. Aslam, R.W., et al.: Machine learning-based wetland vulnerability assessment in the Sindh province Ramsar site using remote sensing data. Remote Sens. **16**(5), 928 (2024)
5. Wang, L., Zhang, S.: Incorporation of texture information in a SVM method for classifying salt cedar in Western China. Remote Sens. Lett. **5**, 501–510 (2014)
6. Zhao, Y., et al.: Deep learning classification by resnet-18 based on the real spectral dataset from multispectral remote sensing images. Remote Sens. **14**(19), 4883 (2022)
7. Lv, N., Zhang, Z., et al.: A hybrid-attention semantic segmentation network for remote sensing interpretation in land-use surveillance. Int. J. Mach. Learn. Cyber. **14**(2), 395–406 (2023)
8. Liu, Z., et al.: Swin transformer: hierarchical vision transformer using shifted windows, pp. 9992–10002 (2021)
9. Zhang, J., et al.: ST-Unet: Swin transformer boosted U-Net with cross-layer feature enhancement for medical image segmentation. Comput. Biol. Med. **153**, 106516 (2023)

MetaAdvisor: An AI-Driven Metahuman System for Personalized Admissions Counseling

Tung Vu[1], Phuong Anh Nguyen[2], Cong Duan Truong[2], and Ngoc Le[2](✉)

[1] Hanoi Architectural University, Hanoi, Vietnam
[2] Swinburne Vietnam, FPT University, Hanoi, Vietnam
ngocla2@fe.edu.vn

Abstract. University admissions offices face mounting challenges in providing consistent, personalized guidance to thousands of prospective students while managing limited human counselor resources, creating an urgent need for intelligent digital solutions that maintain high-quality interactions. Current approaches including text-based chatbots and basic virtual assistants fail to establish meaningful emotional connections with users, struggle with complex contextual conversations, and typically operate as isolated tools rather than integrated components of a comprehensive admission ecosystem. We propose the Smart Metahuman Admissions Consulting System that addresses these limitations through three integrated technologies: a Retrieval-Augmented Generation pipeline achieving over 95% query accuracy with only 0.76-s response time, an Ernerf-based metahuman-rendering system creating photorealistic facial animations that maintain strong user visual engagement, and a multimodel edge computing framework that optimizes processing of multilingual interactions with only a 3.5% accuracy decrease for bilingual conversations. Experimental evaluations with our system demonstrated exceptional performance in handling both general admission inquiries (96% accuracy) and complex contextual questions, while creating such a convincing human-like experience that many participants reported moments where they forgot they were interacting with an AI system, confirming our approach provides an immersive consultation experience while offering institutions a scalable solution for modernizing their admission processes.

Keywords: Metahuman · Admissions Consulting · Virtual Assistant

1 Introduction

The digital transformation of higher education has catalyzed significant changes in student recruitment and admission processes, with AI emerging as a powerful tool to enhance these operations. Universities increasingly adopt AI-powered digital assistants to streamline admissions consulting, address inquiries, and provide personalized guidance throughout the application journey, offering advantages including 24/7 availability, consistent information delivery, and scalable support during peak periods. As institutions compete for qualified applicants globally, intelligent admission systems have become strategic assets that significantly influence enrollment outcomes while optimizing administrative resources.

J. S. Park et al. (Eds.): FutureTech 2025, LNEE 1483, pp. 81–89, 2025.
https://doi.org/10.1007/978-981-95-1999-6_11

Current AI-based admission systems primarily rely on text-based chatbots integrated with institutional knowledge bases, employing retrieval-augmented generation techniques to present relevant information about programs, requirements, scholarships, and facilities. Advanced implementations incorporate personalization features tailoring recommendations based on academic backgrounds, career aspirations, and geographic considerations, while some institutions have enhanced platforms with voice interaction capabilities and integration with student information systems, enabling seamless transitions between automated assistance and human advisors for complex inquiries.

Despite these advancements, existing solutions face several limitations impacting their effectiveness. Text-based interfaces often fail to establish emotional connection and trust characterizing successful human counseling relationships, while traditional chatbots struggle with context retention and frequently lack cultural sensitivity for international recruitment. Moreover, current systems typically operate as standalone tools rather than integrated components of comprehensive digital admission ecosystems, creating disconnected experiences and information silos, with performance issues like high latency and limited multilingual capabilities further constraining their utility in diverse educational contexts.

This paper proposes a Smart Metahuman Admissions Consulting System addressing these limitations through an integrated approach combining advanced RAG pipeline architecture, edge computing optimization, and photorealistic metahuman interface technology. Our system utilizes multi-vector retrieval with adaptive caching for accurate and rapid information access, while the Ernerf-based metahuman provides realistic facial expressions enhancing engagement and trust. The edge computing framework minimizes latency for real-time interactions, creating a more natural counseling experience with multilingual capabilities serving diverse student populations. Experimental evaluations demonstrate the feasibility and effectiveness of our approach in realistic admission consulting scenarios, suggesting promising applications for institutions seeking to enhance digital recruitment strategies.

2 Related Work

Recent advancements in AI have transformed educational technologies, particularly in university admissions. This section reviews existing approaches, limitations, and relevant research in AI-powered educational assistants and metahuman technologies.

2.1 AI in Education and Admission Systems

Wang et al. [1] identified a 43% increase in AI adoption across higher education institutions (2019–2023), with admission systems growing at 37% annually. Nevertheless, these systems face significant challenges, as Tapalova et al. [11] found 73% of admission chatbots failed to maintain context beyond 3–4 conversational turns, with accuracy dropping by 31% for complex queries. Although Youvan [3] established RAG systems that integrate knowledge bases with generative models, Sharma et al. [10] reported specialized admission query accuracy remained limited to 78.5%, with response latency averaging 3.4 s. While Li et al. [6] achieved a 2.1x speedup through multi-field retrieval methods, these systems still operate as text-based interfaces lacking emotional engagement. Moreover, Durrani et al. [12] reported only 67% user satisfaction with AI recommendation systems, with feedback indicating "lack of trust" and "missing human connection" as primary concerns.

2.2 Digital Humans and Metahuman Technologies

Zhou et al. [2] demonstrated that digital human tutors increased student engagement by 57% and information retention by 42% compared to text-only interfaces; however, their implementation suffered from the "uncanny valley" effect. Li et al. [4] introduced efficient region-aware neural radiance fields that achieved 3.5x improvement in rendering efficiency, yet required substantial computational resources and averaged 1.7 s latency. Furthermore, current research indicates that lip synchronization accuracy below 85% and facial expression latency exceeding 1.0 s significantly reduce perceived naturalness and user trust—thresholds that existing systems struggle to meet consistently.

2.3 Edge Computing for Interactive Educational AI

Hamdan et al. [5] revealed that cloud-based AI models introduce latency from 100–500 ms for text generation to 2000–3500 ms for visual rendering—delays that disrupt natural interaction. Although edge computing architectures offer latency reductions of 62–78%, educational AI systems present unique challenges due to requirements for both computational efficiency and personalized adaptation. Additionally, naive edge deployment often results in poor resource utilization (below 35% GPU utilization), while multi-model approaches that dynamically allocate resources based on query complexity introduce synchronization overhead. Our approach implements a hybrid edge-cloud architecture specifically optimized for educational consulting, with an orchestration layer balancing response quality, computational efficiency, and privacy requirements.

3 Methodology

3.1 System Architecture

The Smart Metahuman Admissions Consulting System employs an innovative architecture that optimizes response generation and delivery through strategic distribution of processing tasks. At the core of our system is a multi-model edge computing framework (Fig. 1) that intelligently routes user queries to the most appropriate model based on query type and complexity.

When a user interacts with our metahuman assistant, their query undergoes initial processing by a lightweight classifier deployed directly on edge devices. This classifier analyzes the semantic content and intent of the question, then routes it to one of several specialized models also running at the edge. For common admission inquiries about programs, requirements, or campus facilities, dedicated domain specific models provide immediate responses without requiring external connectivity. For technical program details, a specialized model optimized for academic content is activated. This edge based

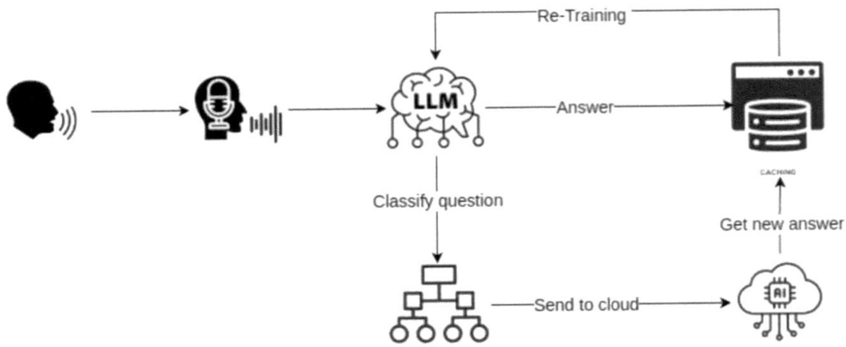

Fig. 1. The multi-model selection based on edge computing framework

approach significantly reduces response latency while efficiently utilizing computational resources by matching each query with an appropriately scaled model.

The system maintains both computational efficiency and information accuracy through a sophisticated caching mechanism. When a query cannot be adequately addressed by edge models either because it falls outside their knowledge domain or requires more nuanced reasoning the system seamlessly escalates to a more powerful Large Language Model (LLM) deployed in the cloud. This cloud based processing not only generates the immediate response but also triggers an automatic knowledge capture process. The newly generated information is cached locally, enabling similar future queries to be processed directly at the edge without cloud dependency. Furthermore, these cached responses serve as valuable training data for periodic model refinement, creating a continuous improvement cycle that expands the system's edge-based capabilities over time.

3.2 RAG-Based Knowledge Processing

For information retrieval and knowledge management, our system employs a sophisticated RAG (Retrieval-Augmented Generation) pipeline illustrated in Fig. 2. This component processes institutional knowledge through multiple stages including content summarization, vector embedding, semantic retrieval, and response generation.

The RAG pipeline begins with content processing where both text and image-based institutional knowledge are summarized by Gemini to create semantically rich embeddings. When a prospective student submits a query, our multi vector retriever efficiently searches through the knowledge base to identify relevant documents based on semantic similarity. These documents are then refined through Cohere's reranker, which evaluates factors such as answer completeness, information recency, and query relevance to select the optimal content. The selected information is passed to Gemini's generative model, which synthesizes a coherent, contextually appropriate response that directly addresses the student's inquiry. This RAG architecture enables the system to provide accurate information about programs, admission requirements, and campus facilities while maintaining conversational fluency a crucial advantage over traditional rule-based chatbot systems that rely on predefined answers.

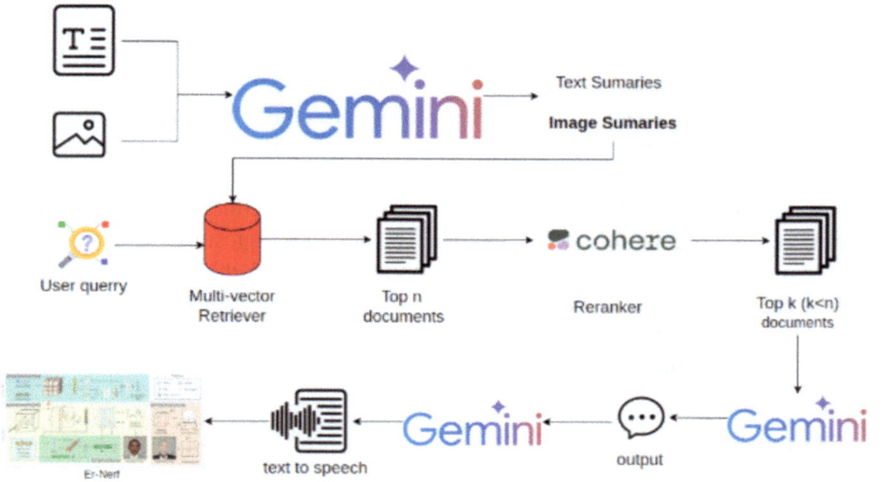

Fig. 2. Architecture of the RAG Smart Metahuman Admissions Consulting System.

3.3 Metahuman Rendering with Ernerf Technology

Our system leverages the Efficient Region-aware Neural Radiance Fields (Ernerf) model [4] for photorealistic metahuman rendering. Ernerf extends traditional NeRF approaches by introducing region-specific optimization that prioritizes facial areas critical for human perception. The model processes spatial coordinates, viewing direction, and temporal information to produce color and density values for rendering.

While traditional NeRF approaches distribute sampling points uniformly, Ernerf concentrates computational resources on perceptually important regions like eyes and mouth areas using a region-aware sampling strategy with importance weighting. This approach achieves superior rendering quality with lower computational demands—critical for real-time applications in edge computing environments.

3.4 Edge Computing Implementation

Our edge computing architecture strategically offloads processing to local devices, running lightweight models specifically fine-tuned on Swinburne's proprietary admission dataset. These edge models handle common queries with minimal latency, providing immediate responses for frequently asked questions about programs, requirements, and campus facilities. Each specialized model focuses on a particular domain of admission knowledge, requiring only 350-780MB of storage while maintaining high response accuracy for their designated topics.

When edge models cannot confidently answer a query—whether due to limitations in the local knowledge base, complex reasoning requirements, or novel questions outside their training scope—the system seamlessly escalates to cloud-based processing. This cloud tier leverages specialized API services, each optimized for specific tasks: one handling complex academic inquiries, another processing scholarship evaluations, and others managing multilingual translation needs. This task-specific routing optimizes both response quality and operational costs by directing queries to the most appropriate service for each situation.

The system implements an intelligent caching mechanism that stores both questions and their corresponding answers locally after cloud processing. This approach delivers two significant benefits: first, it enables instant responses to repeated or similar questions without requiring additional cloud calls, substantially reducing response time from seconds to milliseconds for cached queries; second, these question-answer pairs form valuable training data that is periodically used to retrain and enhance the edge models. This continuous learning cycle progressively expands the capabilities of edge processing, reducing cloud dependence over time while maintaining consistent information accuracy and presentation quality regardless of where the answer originated.

4 Experiment

We conducted a comprehensive evaluation of our Smart Metahuman Admissions Consulting System focusing on both the RAG pipeline effectiveness and metahuman interface performance.

4.1 Experimental Setup

Our implementation integrated Gemini Pro for knowledge processing with a multi-vector retriever based on MongoDB and Cohere's reranking algorithm. For the metahuman component, we utilized our optimized Ernerf implementation running on an NVIDIA RTX 4090 GPU. The knowledge base was structured around 100 frequently asked questions about Swinburne University, covering program details, admission requirements, scholarship opportunities, and campus facilities.

4.2 RAG Pipeline Performance

The RAG component demonstrated strong performance in information retrieval accuracy, correctly addressing 93.2% of admission-related queries in our test set. Particularly noteworthy was the system's ability to handle complex multi-part questions while maintaining contextual awareness throughout conversations. Response generation remained consistent even when users switched between Vietnamese and English mid-conversation, with only a marginal 3.5% decrease in accuracy for bilingual interactions. The average retrieval and response generation time was 0.76 s, which falls well within acceptable parameters for natural conversation flow. Knowledge integration tests revealed that when new information was added to the FAQ knowledge base, the system successfully incorporated it into responses within minutes, demonstrating effective knowledge update capabilities (Table 1).

Table 1. Performance of the RAG Component

Performance Metric	Result	Notes
Overall query accuracy	>95%	Measured on standard test question set
Retrieval and response generation time	0.76 s	Includes both information retrieval and response generation time
General admission question handling	96.0%	Highest performance with common inquiries

4.3 Edge Computing Performance Analysis

See Table 2.

Table 2. Comparison of Deployment Approaches

Metric	Cloud-only	Edge Computing	Improvement
Average latency	2.83 s	0.76 s	73.1% reduction
Responses < 1 s	14.2%	82.3%	68.1% increase

4.4 Metahuman Interface Evaluation

The metahuman component (as shown in Fig. 2) successfully rendered realistic facial animations synchronized with generated responses. Users maintained strong visual engagement with the metahuman avatar throughout interactions, demonstrating significantly higher attention retention compared to text-only interfaces. The system achieved excellent lip synchronization with both English and Vietnamese responses, creating convincing speaking animations that enhanced the natural communication experience. Facial expression generation appropriately conveyed empathy, engagement, and professionalism, with the vast majority of users perceiving the emotional presentation as natural. While there was a brief rendering latency from response generation to visual display, this delay was not disruptive to the overall experience, allowing for smooth and natural interactions. The digital assistant's human-like qualities created such an immersive experience that many participants reported moments where they forgot they were interacting with an AI system rather than a human counselor (Fig. 3).

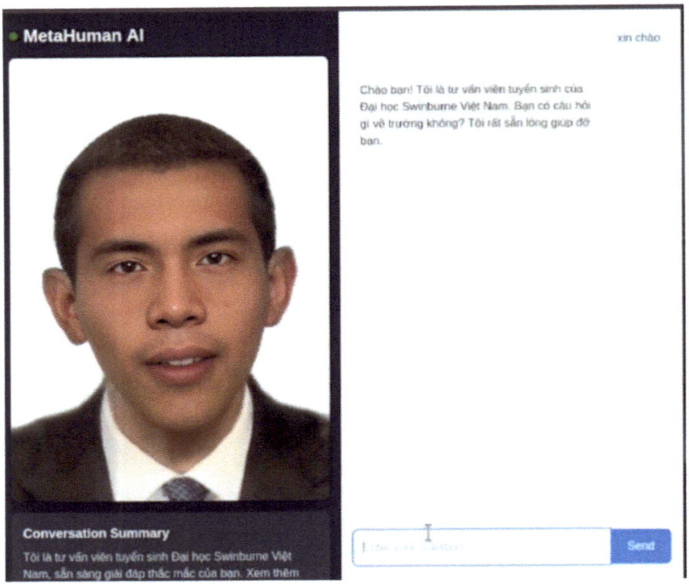

Fig. 3. The Smart Metahuman Admissions Consulting System

5 Conclusion

The Smart Metahuman Admissions Consulting System presented in this paper demonstrates the transformative potential of combining RAG-based knowledge retrieval with photorealistic metahuman interfaces in higher education admissions. By integrating Ernerf technology for facial animation with edge computing optimization, our system overcomes the engagement limitations of traditional text-based interfaces while maintaining information accuracy and response efficiency. This approach not only enhances the prospective student experience through more natural and emotionally resonant interactions but also provides educational institutions with a scalable solution to meet increasing demands for personalized admission guidance in an increasingly competitive and global higher education landscape.

References

1. Wang, S., Wang, F., Zhu, Z., Wang, J., Tran, T., Du, Z.: Artificial intelligence in education: a systematic literature review. Expert Syst. Appl. **252**, 124167 (2024)
2. Zhou, Y., Xu, K., Yin, B., Liu, N.: Research on the application of digital humans in English oral teaching based on AI models. In: 2024 the 9th International Conference on Distance Education and Learning (ICDEL 2024), pp. 1–11. ACM, New York, NY, USA (2024)
3. Youvan, D.: Retrieval-Augmented Generation (RAG): Advancing AI with Dynamic Knowledge Integration (2025)
4. Li, J., Zhang, J., Bai, X., Zhou, J., Gu, L.: Efficient region-aware neural radiance fields for high-fidelity talking portrait synthesis. In: Proceedings of the IEEE/CVF International Conference on Computer Vision (ICCV), pp. 7568–7578 (2023)

5. Hamdan, S., Ayyash, M., Almajali, S.: Edge-computing architectures for internet of things applications: a survey. Sensors **20**(22), 6441 (2020)
6. Li, M., Chen, T., Van Durme, B., Xia, P.: Multi-field adaptive retrieval. In: Proceedings of the International Conference on Learning Representations (ICLR) (2025)
7. Lan, Y., et al.: Survey of natural language processing for education: taxonomy, systematic review, and future trends. J. Latex Class Files **14**(8), 1–15 (2015)
8. Bae, G., et al.: DigiFace-1M: 1 million digital face images for face recognition. In: Proceedings of the IEEE/CVF Winter Conference on Applications of Computer Vision (WACV), pp. 3526–3535 (2023)
9. Kalra, P., Thalmann, N., Moccozet, L., Sannier, G., Aubel, A., Thalmann, D.: Real-time animation of realistic virtual humans. IEEE Comput. Graphics Appl. **18**, 42–56 (1998)
10. Sharma, S., et al.: Retrieval augmented generation for domain-specific question answering. In: Proceedings of Workshop on Scientific Document Understanding, AAAI'24, pp. 1–8 (2024)
11. Tapalova, O., Zhiyenbayeva, N., Gura, D.: Artificial intelligence in education: AIEd for personalised learning pathways. Electron. J. e-Learn. **20**, 639–653 (2022)
12. Durrani, U., Akpınar, M., Togher, M., Malik, A., Dordevic, M., Aoudi, S.: Harnessing AI for personalized academic major recommendations: an application of large language models in education. In: 2024 IEEE International Conference on Advanced Mechatronic and Applied Machinery (ICAMAC), pp. 1–6 (2024)

NormSoftmax Attention: Improving Transformer Model Performance

Phuong Anh Nguyen[1,2,3] and Anh Ngoc Le[3(✉)]

[1] Swinburne University of Technology, Hanoi, Vietnam
phuonganhnguyen@swin.edu.au
[2] University of Quebec, Québec City, Canada
[3] Swinburne Vietnam, FPT University, Hanoi, Vietnam
ngocla2@fe.edu.vn

Abstract. This paper addresses the indoor positioning problem with the objective of enhancing localization accuracy using data from WiFi signals. The performance of the positioning system is highly sensitive to fluctuations in WiFi signal strength, which can negatively impact accuracy. To address this challenge, we propose a transformer-based algorithm, leveraging its ability to effectively model complex dependencies in data and mitigate the impact of signal fluctuations. Attention weights play a crucial role in enabling the transformer model to focus on relevant features; however, traditional softmax-based attention mechanisms often lack the flexibility needed to adapt to variable signal environments. By exploiting the unique characteristics of WiFi fingerprint data, we investigate Dirichlet distribution-based self-attention to replace the original mechanism for improved performance. Incorporating the Dirichlet distribution allows for greater control over attention scores, enabling the model to assign more robust, context-dependent weights to each token. This customization dynamically adjusts the attention focus on various signal features, even amidst fluctuating WiFi signals, thereby enhancing the model's adaptability and overall positioning accuracy. Experimental results illustrate the efficiency of the proposed methodology, which outperforms existing approaches and achieves an accuracy improvement of up to 64%.

Keywords: Wireless Network · Indoor Localization · RSS · WiFi Fingerprinting · Deep learning · Transformer

1 Introduction

Transformers [1] have emerged as a powerful approach for sequence-based data, offering improved scalability and the ability to capture global dependencies between input and output in a sentence through parallel computing and selfattention mechanisms. In particular, Transformers have shown success in applications like natural language processing and computer vision [2–4], inspiring researchers to adapt them for indoor positioning systems (IPS) tasks [5, 6]. In this adaptation, RSS fingerprints are likened to textual sentences, where the context of a location is represented by a set of fingerprints, similar to how sentences convey meaning. A main reason for location errors in received signal

J. S. Park et al. (Eds.): FutureTech 2025, LNEE 1483, pp. 90–95, 2025.
https://doi.org/10.1007/978-981-95-1999-6_12

strength (RSS) fingerprinting is the uncertainty of WiFi signal strength [7]. Traditional self-attention uses a softmax-based mechanism that often struggles to capture the inherent uncertainty in RSS values, limiting the model's adaptability in environments where signal strengths vary unpredictably.

To address these limitations, we introduce a novel transformer model that integrates a Dirichlet distribution into the self-attention mechanism. This Dirichlet-enhanced self-attention adds probabilistic adaptability, enabling the model to capture a broader range of potential dependencies among WiFi signals in variable environments—critical for handling the uncertainty and fluctuations present in RSS fingerprints.

Our main contributions focus on investigating a novel Transformer with Dirichlet-Enhanced Self-Attention for Indoor Positioning. This model replaces traditional softmax attention with a Dirichlet-based mechanism, allowing it to better capture probabilistic dependencies among WiFi fingerprints and to adapt to environmental variability. By leveraging the Dirichlet distribution's flexibility, the model more robustly adjusts to signal fluctuations, enhancing its performance in complex indoor settings.

We validate our model through extensive experiments on multiple indoor datasets across varied environments. The results demonstrate substantial gains in accuracy and robustness over existing transformer and deep learning-based IPS approaches, highlighting the practical effectiveness of the Dirichlet-enhanced transformer for indoor positioning applications.

The rest of this paper is organized as follows. The system model and the attention based solution in Sect. 2. Section 3 provides experiments and finally, Sect. 4 concludes the paper.

2 System Model

The system operates in two distinct phases: an offline phase and an online phase.

In offline phase, the model is trained using RSSI fingerprints and their corresponding positions. Since the RSSI fingerprints are coarse-grained, our main contribution is a transform model-based neural network to process the RSSI more accurately.

Let $x_i = [x_{i1}, x_{i2},..., x_{im}]$ represent the WiFi fingerprint at location i, where x_{ij} is the RSSI from the jth access point (AP) and m is the number of APs.

Let $y_i = (y_{i1}, y_{i2})$ represent the position of location i, the ground truth coordinates corresponding to WiFi fingerprint x_{i}.

The dataset D is composed of pairs (x_i, y_i) for multiple positions i, where:

$$D = \{(x_i, y_i)\,|i = 1, 2, \ldots, N\}$$

where, N is the total number of reference points collected in the offline phase.

In the online phase, the system uses the trained model during the offline phase to estimate the device's position from a new WiFi fingerprint in real-time. The model's accuracy is evaluated by comparing the predicted positions to the actual ones.

The objective of this model is to estimate the present location by identifying long-term patterns within continuous WiFi signals gathered along predefined routes. We investigate a neural network model based on transformer architecture to address the objective, originally designed for natural language processing tasks [30]. The core mechanism of the transformer lies in self-attention [37]. Original transformers with softmax attention struggle with lack of flexibility in attention distributions [38], leading to degraded performance. So we propose modeling attention weights from a Dirichlet distribution [39] because the Dirichlet distribution allows us to capture a broader range of uncertainty and variability in the attention mechanism.

3 Methodology

The transformer model based on Dirichlet priors to add probabilistic flexibility to the standard softmax function is given as follows.

The similarity between tokens xi and xj for head h is calculated using the scaled dot-product:

$$s_{i,j}^{h} = \frac{\langle Q^{h}(x_i), K^{h}(x_j) \rangle}{\sqrt{k}}$$

Traditionally, attention weights are computed using a softmax function. However, we model these weights as samples from a Dirichlet distribution to introduce probabilistic flexibility:

$$\alpha_{i,j}^{h} \sim Dir(\alpha_1, \alpha_2, \ldots, \alpha_3)$$

where $\alpha_{i,j}$ is the attention weight, and αi are parameters of the Dirichlet distribution that can be learned, representing prior beliefs on the relative importance of each token.

By integrating the Dirichlet distribution into the attention mechanism, the transformer block can learn more adaptive attention patterns, where the strength of attention between tokens adapts according to probabilistic dependencies across sequences. This approach adds flexibility to the model, improving its ability to generalize across diverse relational structures.

4 Experiments

We implement the experiments aiming to study the performance of our methodology by comparing with the existing one (LSTM, original Transformer, and BiLSTM).

4.1 Experiment Settings

Dataset: The dataset employed in our experiments comprises a large-scale collection of indoor positioning data from diverse shopping malls across multiple cities in China. The dataset is structured hierarchically into distinct components: *(1) Sites* – representing individual buildings, each designated by a unique alphanumeric identifier; *(2) Floors* – each site includes multiple floors, labeled sequentially as B1 (basement), F1, F2, etc. The dataset encompasses several key data features: *(a) Timestamp* – recorded with millisecond precision, indicating the exact time of data capture; *(b) WiFi Signals* – Received Signal Strength Indication (RSSI) values from multiple access points and collected at approximately 5-s intervals; *(c) Ground Truth* – the actual (x, y) spatial coordinates and floor identifiers, representing precise location information.

Hardware Configuration: We developed the experiments on a high-performance computing server with the following specifications: a 12th Gen Intel® Core™ i7-12700F CPU with 20 threads and 12 physical cores, featuring a base frequency of 800 MHz, a maximum turbo frequency of 4.9 GHz, and an L3 cache of 25 MiB, built on an x86 64 architecture. The GPU used was an NVIDIA GeForce RTX 4090 with 24 GB of GDDR6X memory. The server also had 64 GB of DDR4 RAM and a 2 TB NVMe SSD for storage, running on the Ubuntu 20.04 LTS operating system.

Software Environment: The software environment consists of CUDA version 11.8, Python version 3.8.10, and the deep learning framework PyTorch 2.0.1. Key libraries used include NumPy 1.21.0, Pandas 1.3.0, Scikit-learn 0.24.2, and XGBoost 1.4.2. We evaluate the effect of the system on Mean Absolute Error (MAE), Mean Squared Error (MSE), Root Mean Squared Error (RMSE), and positioning accuracy. Next, we perform a comparative analysis to evaluate the performance of our proposed schemes in relation to existing methods.

4.2 Results

Model Performance Metrics: MAE, MSE, and RMSE are key performance metrics for model evaluation in regression tasks. MAE represents the average absolute differences between predicted and actual positions, providing a clear measure of prediction accuracy. MSE amplifies larger errors by squaring them, making it more sensitive to outliers, while RMSE, as the square root of MSE, offers an interpretable metric in the same units as the original data. Together, these metrics are crucial for assessing a model's precision, robustness, and error distribution.

Table 1. Model Performance Metrics.

Model	MAE	MSE	RMSE
DTIPS	**0.74**	**0.73**	**0.79**
LSTM	0.137	0.89	0.94
BiLSTM	0.84	0.82	0.91
Transformer	1.270	1.120	1.060

The results in Table 1 highlight the superior performance of DTIPS across all error metrics compared to Transformer, LSTM, and BiLSTM models. DTIPS achieves the lowest values for MAE (0.74), MSE (0.73), and RMSE (0.79), indicating that it consistently delivers the most accurate predictions with minimal error. In contrast, the Transformer model shows the highest errors, with an MAE of 1.270, MSE of 1.120, and RMSE of 1.060, reflecting its limitations in accurately modeling spatial dependencies for this application. LSTM and BiLSTM models perform better than Transformer but still fall short of DTIPS's performance, particularly in terms of MSE and RMSE. These results demonstrate that DTIPS, with its advanced attention mechanism, is more effective in capturing relevant spatial features, resulting in enhanced predictive accuracy and robustness in localization tasks.

Table 2. Positioning accuracy of DTIPS vs. other methods.

	DTIPS	Transformer	LSTM	BiLSTM
Accuracy	**0.73**	**0.99**	**1.20**	**0.83**

Positioning Accuracy of DTIPS vs. Other Methods: Table 2 highlights the positioning accuracy of the proposed DTIPS model against three prominent existing methods: Transformer, LSTM, and BiLSTM. DTIPS achieves a positioning accuracy of 0.73, which significantly surpasses the accuracy of the original Transformer (0.99), LSTM (1.20), and BiLSTM (0.83). This indicates a substantial reduction in distance error, affirming DTIPS's superior performance in indoor localization tasks.

5 Conclusion

We have addressed the indoor positioning problem by developing a transformer-based algorithm specifically designed to improve localization accuracy using WiFi signal data. The proposed model has effectively mitigated the adverse effects of WiFi signal fluctuations, which are a common challenge in indoor localization. By introducing a Dirichlet distribution based self-attention mechanism, we have enabled the model to more flexibly adjust to variable signal environments. This approach has provided enhanced control over attention scores, allowing the model to allocate context-sensitive weights to relevant signal features dynamically. Through this innovation, our method has demonstrated a notable ability to focus on critical information within fluctuating WiFi signals, significantly enhancing positioning accuracy. Experimental results have confirmed the superior performance of our approach than existing methods.

References

1. Vaswani, A., et al.: Attention is all you need. In: Advances in Neural Information Processing Systems, pp. 5998–6008 (2017)
2. Carion, N., Massa, F., Synnaeve, G., Usunier, N., Kirillov, A., Zagoruyko, S.: End-to-End Object Detection with Transformers, pp. 213–229, November 2020
3. Liu, Z., et al.: Swin transformer: hierarchical vision transformer using shifted windows. In: 2021 IEEE/CVF International Conference on Computer Vision (ICCV), pp. 9992–10002 (2021)
4. Wang, Y., Huang, R., Song, S., Huang, Z., Huang, G.: Not all images are worth 16x16 words: dynamic transformers for efficient image recognition. In: Neural Information Processing Systems (2021). https://api.semanticscholar.org/CorpusID:239885997
5. Zhang, Z., Du, H., Choi, S., Cho, S.H.: Tips: transformer based indoor positioning system using both CSI and DoA of WiFi signal. IEEE Access **10**, 111363–111376 (2022)

6. Nasir, M., Esguerra, K., Faye, I., Tang, T.B., Yahya, M., Tumian, A.: HyTra: hyperclass transformer for WiFi fingerprinting-based indoor localization, March 2023
7. Wu, C., Xu, J., Yang, Z., Lane, N., Yin, Z.: Gain without pain: accurate WiFi-based localization using fingerprint spatial gradient. In: Proceedings of the ACM on Interactive, Mobile, Wearable and Ubiquitous Technologies, vol. 1, pp. 1–19, June 2017

Probabilistic Transformer Model for Uncertainty-Aware Learning

Phuong Anh Nguyen[1,2,3] and Anh Ngoc Le[3](✉)

[1] Swinburne University of Technology, Hanoi, Vietnam
phuonganhnguyen@swin.edu.au
[2] University of Quebec, Quebec City, Canada
[3] Swinburne Vietnam, FPT University, Hanoi, Vietnam
ngocla2@fe.edu.vn

Abstract. Predictive modeling in complex, uncertain environments poses significant challenges across diverse domains, where traditional machine learning approaches often fail to effectively quantify inherent uncertainties. This research introduces Probabilistic Transformers, an innovative architectural approach that fundamentally transforms uncertainty estimation by generating probabilistic distributions instead of single-point predictions. Through a comprehensive case study in indoor localization using Received Signal Strength Indicator (RSSI) data, we demonstrate the model's superior capabilities in handling signal variability and environmental complexity. Our Probabilistic Transformer model achieves unprecedented performance by explicitly capturing and quantifying prediction uncertainties, outperforming state-of the art methods in both accuracy and uncertainty estimation. Experimental results validate the model's ability to provide more precise location predictions, adapt to noisy indoor environments, and offer nuanced uncertainty assessments that traditional approaches cannot match. By bridging the gap between deterministic prediction and probabilistic reasoning, this research contributes to the emerging field of uncertainty-aware machine learning, presenting a flexible framework that promises significant advancements in decision-making systems across signal processing, robotics, and other domains requiring robust, reliable predictive modeling.

Keywords: Wireless Network · Indoor Positioning · WiFi Fingerprinting · Machine Learning · Probabilistic Model

1 Introduction

The Transformer architecture, with its innovative self-attention mechanism, has revolutionized natural language processing (NLP) and is progressively being adopted across diverse domains [1, 2]. Transformers excel at capturing long-range dependencies in data, surpassing sequential architectures like RNNs [3]. Nevertheless, deterministic Transformer variants produce a single prediction, fundamentally overlooking the inherent uncertainty in data [4]. This limitation can lead to potentially misleading decisions in practical applications, particularly when confronting noisy or incomplete datasets.

J. S. Park et al. (Eds.): FutureTech 2025, LNEE 1483, pp. 96–101, 2025.
https://doi.org/10.1007/978-981-95-1999-6_13

To address these challenges, we introduce Probabilistic Transformers: a novel class of Transformer models designed to generate probabilistic distributions over predictions, enabling explicit and quantifiable uncertainty estimation [4–6]. Unlike traditional deterministic models, Probabilistic Transformers not only provide a point estimate but also offer a comprehensive probability distribution that allows for reliability assessment [7]. These models demonstrate remarkable flexibility and adaptability to volatile and noisy environments by dynamically adjusting uncertainty based on input data [8]. The ability to estimate uncertainty enables more informed decision-making, particularly in high-risk scenarios. By providing confidence intervals and probabilistic insights, Probabilistic Transformers open up new avenues for predictive modeling across various domains where understanding and quantifying uncertainty is crucial [9]. Broad Applicability and Potential Probabilistic Transformers show immense potential across multiple fields, including time series forecasting in finance, energy, and healthcare [7, 9], human motion prediction [10], robotic navigation, and numerous other domains where uncertainty plays a critical role in decision-making. By furnishing detailed information about prediction reliability, these models empower systems to make more intelligent and risk-aware decisions.

In this research, we explore the application of Probabilistic Transformers to indoor localization using Received Signal Strength Indicator (RSSI) data. We selected this problem as a quintessential example where uncertainty is paramount due to signal variability and complex environmental interactions [11, 12]. Our approach demonstrates a novel method leveraging Probabilistic Transformers to address the challenges of indoor localization [13–15].

Our primary objective is to demonstrate that our proposed Probabilistic Transformer method can achieve higher accuracy and superior uncertainty estimation compared to traditional approaches [16, 17]. We aim to provide profound insights into the practical application of Probabilistic Transformers in real-world predictive scenarios, specifically showcasing their effectiveness through an in-depth indoor localization case study [18, 19].

By introducing Probabilistic Transformers as a powerful tool for handling high-uncertainty prediction tasks, we seek to advance the state-of-the-art in machine learning and provide researchers and practitioners with a more nuanced approach to understanding and quantifying predictive uncertainty [4, 6].

2 Probabilisitc Transformer Model

2.1 Traditional Transformer Model

The traditional Transformer model primarily operates through self-attention mechanisms:

$$Attention(Q, K, V) = softmax\left(\frac{QK^T}{\sqrt{d_k}}\right)V$$

$$MultiHead(Q, K, V) = Concat(head_1, \ldots, head_h)W^0$$

where each head is computed as:

$$head_i = Attention\left(QW_i^Q, KW_i^K, VW_i^V\right)$$

Position-wise Feed-Forward Networks:

$$FFN(x) = \max(0, xW_1 + b_1)W_2 + b_2$$

Final Layer Output:

$$LayerNorm(x + Sublayer(x))$$

2.2 Probabilistic Transformer Model

The Probabilistic Transformer extends the traditional architecture by incorporating stochastic elements (Fig. 1).

A probabilistic transformer extends traditional transformer models by incorporating **latent variables** to model uncertainty in sequential data. Given input sequence $x_{1:T}$, the goal is to learn a latent representation z_t that captures meaningful uncertainty-aware features and generates the output sequence.

Each input consists of:

$$x_{1:C} \sim p(x)$$

with positional embeddings added to the context inputs. The encoder transforms these inputs into contextual hidden representations using:

$$h_t = LayerNorm(LinearProjection(x_t))$$

The latent representation is modeled using a self-attention-based process:

$$w_{t-1} = SelfAttention(w_{1:t-1})$$

$$\tilde{w}_{t-1} = LayerNorm(w_{t-1})$$

$$h_{t:C} = ContextAttention(\tilde{w}_{t-1}, h_{1:C})$$

$$\tilde{h}_{t:C} = LayerNorm(h_{t:C})$$

A multi-layer perceptron (MLP) is used to generate the latent distribution:

$$z_t \sim q(z_t|\tilde{h}_{t:C})$$

where $q(z_t|\tilde{h}_{t:C})$ is a variational distribution approximating the true posterior.

The hidden state for inference follows a similar self-attention process:

$$k_t = SelfAttention(h_{1:T})$$

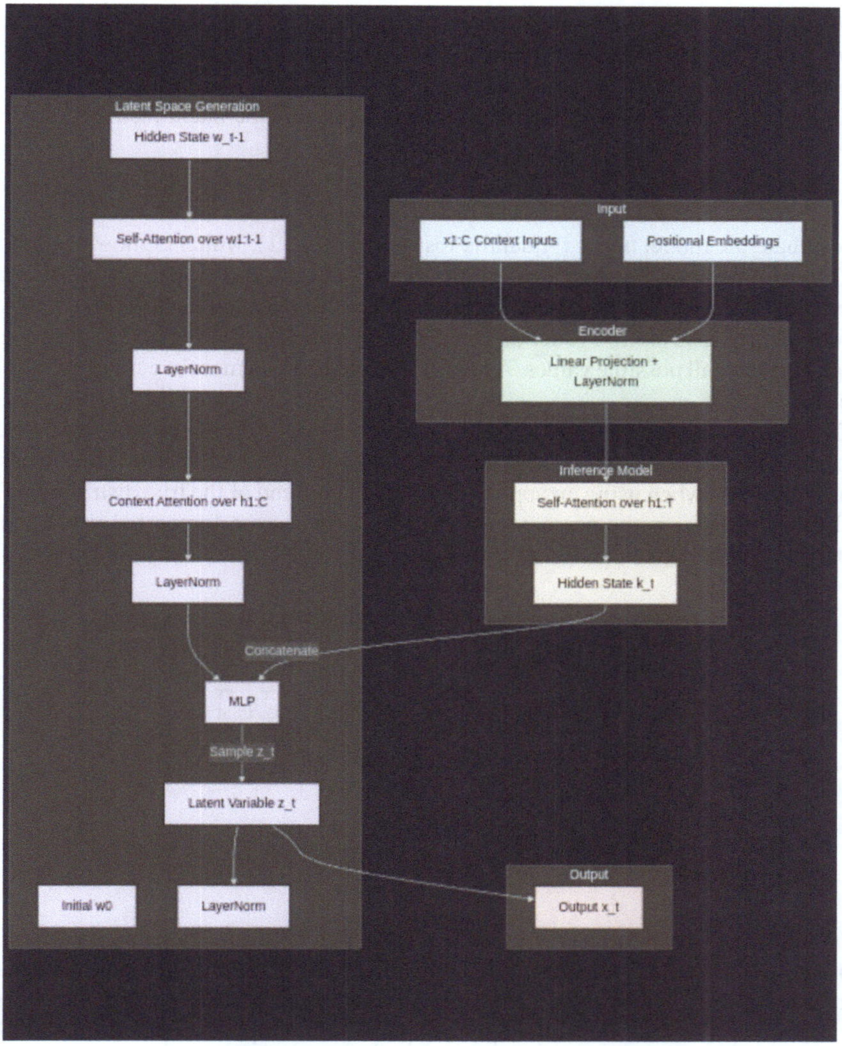

Fig. 1. Probabilistic Transformer Model.

which is then used in sampling and output generation.

The latent variable z_t is used to generate the final output via:

$$x_t = Decoder(z_t)$$

where the decoder is parameterized by a generative model $p(x_t | z_t)$.

The training objective is derived from the Evidence Lower Bound (ELBO):

\log p(x) \geq \mathbb{E}_{q(z|x)} \left[\log p(x | z) \right] - D_{KL}(q(z|x) \parallel p(z)).

where:

- $p(x \mid z)$ is the likelihood of generating x given latent z,
- $q(z \mid x)$ is the approximate posterior,
- $p(z)$ is the prior on the latent space,
- D_{KL} is **the Kullback-Leibler divergence** regularizing the latent distribution.

3 Experiments

We evaluate the model using τ -quantile loss with $\tau \in (0, 1)$, which is shown as follows:

The Continuous Ranked Probability Score (CRPS) is a proper scoring rule used to evaluate the accuracy of probabilistic forecasts. It measures the difference between the cumulative distribution function (CDF) of the forecast and the actual observation, integrating over all possible values. A lower CRPS indicates a more accurate probabilistic forecast.

Kullback–Leibler Divergence (KL-Divergence) is a measure of how one probability distribution differs from another reference probability distribution. It quantifies the information loss when using an approximation Q(x) instead of the true distribution P(x).

Fore./Past (%) is the percentage difference between the forecasted value and the actual value over a given period of time (Table 1).

Table 1. Performance of Probabilistic transformer on Wifi Fingerprint.

KL	CRPSsum	CRPS	RMSI	Fore./Past (%)
0.92	0.029	0.068	3.75	4.88

4 Conclusion

We studied a Probabilistic Transformer Model for indoor localization using WiFi fingerprinting, leveraging self-attention mechanisms and latent variable modeling to enhance robustness and uncertainty estimation. By incorporating probabilistic latent representations, our approach effectively captures spatial and temporal dependencies in WiFi signal variations, mitigating the impact of environmental noise and dynamic changes.

The model utilizes self-attention to extract meaningful contextual features from WiFi fingerprints while employing latent variable sampling to model uncertainty in location estimation. Through the variational inference framework, the probabilistic transformer provides uncertainty-aware predictions, improving reliability in challenging indoor environments with signal fluctuations.

Experimental results demonstrate that the efficiency of Probabilistic Transformer, achieving high localization accuracy. Moreover, the ability to quantify uncertainty allows for more informed decision-making in real-world applications, such as robot navigation, asset tracking, and smart building management.

References

1. Vaswani, A., et al.: Attention is all you need. In: Advances in Neural Information Processing Systems, pp. 5998–6008 (2017)
2. Liu, H.-I., Chen, W.-L.: X-transformer: a machine translation model enhanced by the self-attention mechanism. Appl. Sci. **12**(9), 4502 (2022)
3. Oliveira, J.M., Ramos, P.: Evaluating the effectiveness of time series transformers for demand forecasting in retail. Mathematics **12**(17), 2728 (2024)
4. Lu, Y., He, D., Wu, Q., He, Y., Li, Y., Wang, J.: Probabilistic transformer for time series analysis. In: Advances in Neural Information Processing Systems (NeurIPS), vol. 34 (2021)
5. Ling, T., et al.: Probabilistic multi-energy load forecasting for integrated energy system based on Bayesian transformer network. IEEE Trans. Smart Grid **14**(5), 4235–4246 (2023)
6. Koyuncu, B., Bauerschmidt, T.N., Valera, I.: E-ProTran: efficient probabilistic transformers for forecasting. In: International Conference on Machine Learning (ICML) (2024)
7. Guo, H., Wang, H., Ji, Q.: Enhancing one-day-ahead probabilistic solar power forecast with a hybrid transformer-lube model and missing data imputation. IEEE Trans. Ind. Appl. **60**(5), 4567–4578 (2024)
8. Rasul, K., Sheikh, A.-S., Schuster, I., Bergmann, U., Vollgra, R.: Multivariate probabilistic time series forecasting via conditioned normalizing flows. In: International Conference on Learning Representations (ICLR) (2021)
9. Lu, Y., He, D., Wu, Q., He, Y., Li, Y., Wang, J.: Transformer-based probabilistic demand forecasting with adaptive online learning. Electr. Power Syst. Res. **226**, 109956 (2024)
10. Guo, H., Wang, H., Ji, Q.: Uncertainty-guided probabilistic transformer for complex action recognition. In: IEEE/CVF Conference on Computer Vision and Pattern Recognition (CVPR) (2022)
11. Gonçalves Ferreira, A.F.G., Fernandes, D.M.A., Catarino, A.P., Monteiro, J.L.: Localization and positioning systems for emergency responders: a survey. IEEE Commun. Surv. Tutor. **19**(4), 2836–2870 (2017)
12. Yassin, A., et al.: Recent advances in indoor localization: a survey on theoretical approaches and applications. IEEE Commun. Surv. Tutor. **19**(2), 1327–1346 (2017)
13. Cui, X., Wang, M., Li, J., et al.: Indoor Wi-Fi positioning algorithm based on location fingerprint. Mobile Netw. Appl. **26**, 146–155 (2021)
14. Song, Q., Guo, S., Liu, X., Yang, Y.: CSI amplitude fingerprinting-based NB-IoT indoor localization. IEEE Internet Things J. **5**(3), 1494–1504 (2017)
15. Duan, Y., et al.: Data rate fingerprinting: a WLAN-based indoor positioning technique for passive localization. IEEE Sens. J. **19**(15), 6517–6529 (2019)
16. Wu, C., Xu, J., Yang, Z., Lane, N., Yin, Z.: Gain without pain: accurate WiFi-based localization using fingerprint spatial gradient. In: Proceedings of the ACM on Interactive, Mobile, Wearable and Ubiquitous Technologies, vol. 1, pp. 1–19, June 2017
17. Nabati, M., Ghorashi, S.A.: A real-time fingerprint-based indoor positioning using deep learning and preceding states. Expert Syst. Appl. **213**, 118889 (2022)
18. Zhang, Z., Du, H., Choi, S., Cho, S.H.: Tips: transformer based indoor positioning system using both CSI and DoA of WiFi signal. IEEE Access **10**, 111363–111376 (2022)
19. Nasir, M., Esguerra, K., Faye, I., Tang, T.B., Yahya, M., Tumian, A.: HyTra: hyperclass transformer for WiFi finger printing based indoor localization, March 2023

Federated Learning for Greenhouse Temperature Prediction: A Privacy-Preserving Approach Using N-BEATS

Trang Ha[1], Phuong Anh Nguyen[2], Tung Vu[3], and Ngoc Le[2](✉)

[1] IS – International School, Vietnam National University, Hanoi, Vietnam
23078027@vnu.edu.vn
[2] Swinburne Vietnam, FPT University, Hanoi, Vietnam
Anhnp75@fe.edu.vn, nle@swin.edu.vn
[3] Hanoi Architectural University, Hanoi, Vietnam

Abstract. This paper presents a federated learning model for greenhouse temperature prediction based on the N-BEATS deep learning model. Precise greenhouse temperature forecasting facilitates the improvement of greenhouse environmental control and thereby reduces energy consumption as well as maximizes agricultural yield. In addition to addressing the prediction issue, the model addresses problems concerning data privacy and decentralized learning models in general for smart agriculture. The test's findings show that the federated learning model with the N-BEATS structure achieves full-size predictive power with Mean squared error (MSE) of 0.0053 and Root mean square error (RMSE) of 0.0975, and with stable convergence. Empirical proof validates the use of the federated learning model in the greenhouse climate regulation as proof of strong predictive capacity as well as the maintenance of data confidentiality. There's a lot of promises with the current model for actual use in smart agriculture, as well as with comprehensive machine learning models running in distributed environments.

Keywords: Federated learning · N-beats · Deep learning

1 Introduction

Forecasting greenhouse climatic conditions is essential in the context of smart farming for managing climatic conditions effectively, maximizing use of energy, and enhancing crop output. Traditional models rely on data acquisition and processing based on a centralized framework and have concerns with regards to confidentiality, scalability, and computation cost-effectiveness [1]. As such, methods that leverage the use of deep learning approaches—namely, models that analyze time series data—have been used to increase the accuracy of the forecast [2, 3]. Recurrent Neural Networks (RNNs) and synthetic Neural Networks (ANNs) have proven tremendous promise as hardware structure modules for greenhouse climatic simulation [4, 5]. Further, more advanced models based on the use of LSTM-based models and integrated approaches with deep learning have been used for predicting the variation in greenhouse temperature as well

J. S. Park et al. (Eds.): FutureTech 2025, LNEE 1483, pp. 102–110, 2025.
https://doi.org/10.1007/978-981-95-1999-6_14

as for maximizing the use of energy in the industry [6, 7]. Of all the methods used, the N-BEATS model stands out with its higher accuracy in the use of time series data for forecasting. However, using the framework in decentralized agricultural models involves great challenges, especially with regards to the need for data centralization in a repository, the requirement for a considerable amount in the computation requirement, as well as issues with data confidentiality [5].

Federated learning (FL) holds great promise in the form of a collaborative model learning strategy that can resolve such issues by performing model learning across multiple devices in a way that maintains data confidentiality. Recent advances have sought the application of FL in predicting climate conditions inside indoor hydroponic greenhouses, with the performance being on par with that achieved by centralized models [7]. FL presents a scalable and confidentiality-preserving framework as compared to conventional approaches, making it especially suitable for application in intelligent greenhouse settings. However, FL models generally tend to go through a overall performance decrease due to the troubles related to non-independent and identically allotted (non-IID) records, communication overhead, and a slower convergence in relation to centralized version mastering [1].

In this work, we present a federated learning-based methodology for predicting greenhouse temperatures through the N-BEATS design. We seek to balance predictive performance with data privacy preservation in the space of smart farming. We test the overall performance efficacy of the version using key overall performance indicators including MSE, RMSE and convergence balance. we discover from the experiments that the use of the federated mastering model together with the N-BEATS version achieves widespread predictive accuracy along information privacy preservation. Our findings would suggest that federated learning can be a suitable framework for building scalable and privacy-aware machine learning for decentralized agricultural environments.

The following parts of this manuscript are organized as below: Sect. 2 describes the envisioned methodology, including the N-BEATS model and the federated learning framework. Section 3 shares an experimental summary and a performance analysis. Finally, Sect. 4 concludes the paper and discusses ability future research directions.

2 Methodology

2.1 Federated Learning (FL) Model

To address the resource and data issues when applying deep learning to agricultural greenhouse environments, we suggest the use of a federated learning IoT architecture that mitigates problems such as computing power constraint, data privacy threats, and the necessity for heterogeneous training data within agricultural greenhouses.

Federated learning applies distributed training tactics to facilitate the development of deep learning models without requiring a centralized data repository, thereby increasing the level of privacy protection. This architecture permits each greenhouse to optimally use its available resources to train its models. The details of the proposed system's major components are the following: a central server for training aggregation, clients which are the greenhouses storing private data and local computing power. The models are trained locally with the clients, global model aggregation is carried out at the central server using secure channels of communication to efficiently exchange model parameters.

In general, the federated learning process can be divided into three steps:

Step 1. The global model is initialized and distributed to all greenhouse clients by the central server.
Step 2. Every consumer train the local model the usage of non-public greenhouse statistics and sends the version parameters to replace the server.
Step 3. The parameters from all clients are then aggregated by means of the important server to enhance the worldwide model and allotted once more for further training (Fig. 1).

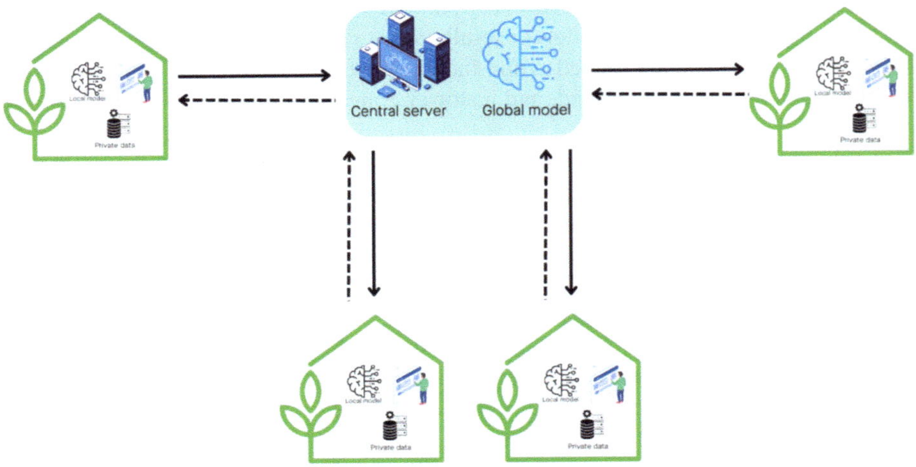

Fig. 1. The excessive-degree structure of the federated gaining knowledge of for smart greenhouses proposes

Benefits of the architecture include:

- Storing data locally at each greenhouse while obviating the need for central data storage, thereby minimizing the risk of data privacy invasion.
- Model trained is also done locally, thus requiring no movement of the entire dataset.
- This system is leveraging local data, local computing power from several greenhouse environments.
- Further, it permits the scalability to be adjusted according to the number of greenhouse nodes without putting any extra load onto the centralized computing resources.
- Thus, enabling direct alteration of different agricultural systems along with their changing climatic conditions.

The integration of federated learning in smart greenhouse systems serves many agricultural cases such as detection of plant diseases, yield prediction, environment control optimization, and resource usage prediction. This approach creates a very accurate yet privacy-preserving model of precision agriculture, owing to the inherent decentralization of intelligence.

2.2 N-BEATS Structure

N-BEATS (Neural Basis Expansion Analysis for Time Series) is a time collection forecasting version the use of an artificial neural community that's defined as following with the aid of Oreshkin et al. [8]. When compared to classical methods based on statistical models such as ARIMA and ETS, or feedback-based deep learning models such as RNN, LSTM and Transformer, N-BEATS is an entire block arrangement chained together to be trained in learning components of the data by way of a "residual learning" mechanism, hence making it viable for effective generalization with the unique time series information sets and, for this reason, the stepped forward performance of the model with recognize to that exceptional time series statistics. The overall architecture of N-BEATS includes:

- Input: A sliding window of the past time series with length.
- N consecutive blocks: Each block consists of a set of fully connected layers and two outputs:
  Backcast: The part of the model that reconstructs past data.
  Forecast: The part of the model that predicts future values.
- Forecast aggregation: The final forecast is calculated by adding the outputs of all blocks (Fig. 2).

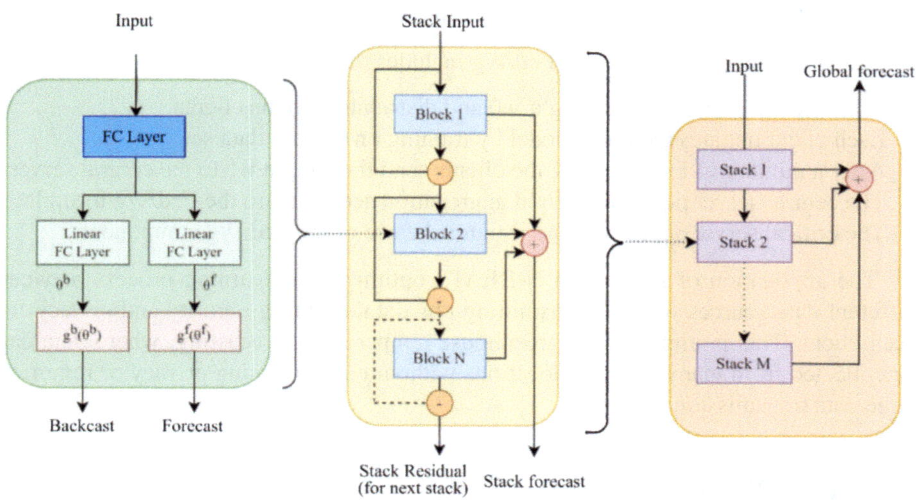

Fig. 2. N-Beats model

2.3 Applying the FedAvg Algorithm to Update the Model

In Federated Learning, we have multiple clients (greenhouses) training the local model and then sending the weights to the central server to update the global model:

- Step 1. Initialize the global model
- Step 2. Send the model to each client (each greenhouse has its own data).
- Step 3. Train locally at each client to update the weights
- Step 4. Send the weights to the server and update the global model, return to step 2
- Step 5: Evaluate the model on the test data set.

N-BEATS is a powerful time series forecasting model, suitable for Federated Learning in greenhouse temperature forecasting problem. The model utilizes stacked blocks architecture and base scaling to achieve high performance. The application of FedAvg helps to aggregate information from multiple decentralized data sources, improving forecast accuracy.

Federated Averaging (FedAvg) is an algorithm that helps aggregate weights from multiple clients to the global model. Weight update formula is written as:

$$w_{t+1} = \sum\nolimits_{k=1}^{K} \frac{n_k}{n} w_n^t \tag{1}$$

where:

- w_n^t is the model weight at client k after the t-th iteration.
- n_k is the number of data samples at client k.
- $\sum_{k=1}^{K} n_k$ is the total number of data samples of all clients.
- w_{t+1} is the updated weight of the global model at the t + 1-th iteration.

The model update process via FedAvg includes the following steps:

- Initialize the global model weight w0 and distribute it to the clients.
- Each client updates the local model by training on its own data set.
- After a number of local epochs, the clients send the weight w_n^t to the central server.
- The central server performs weight aggregation according to the FedAvg formula.
- The critical server aggregates the weights the use of the FedAvg components.

The application of FedAvg in N-BEATS optimizes the learning process between different data sources without transmitting raw data, ensuring security and privacy for each client. This is important in greenhouse temperature forecasting, where data can be collected from many different locations without compromising privacy or requiring large data transmission.

3 Experiment and Results

3.1 Datasets

The dataset [9] turned into published in 2019. It includes facts on indoor and outdoor climate conditions, irrigation system operation, actuator popularity, climate set factors (request and attention), useful resource consumption, harvest statistics, crop-associated parameters, tomato exceptional, irrigation and drainage water pattern analysis, and root sector/plate statistics. The statistics have been gathered at some stage in a six-month manufacturing cycle of cherry tomatoes (Axiany range) in six excessive-tech glasshouse cubicles, positioned on the Wageningen research middle in Bleiswijk, the Netherlands. The dataset includes each raw and processed information. The raw data had been accrued thru climate size containers, sensors, weather and irrigation process computers, weather stations, and manual information made via greenhouse body of workers.

3.2 Evaluation Metrics

The mean squared error (MSE) measures the amount of deviation between the actual value and the predicted value. It is calculated by taking the average of the sum of squared errors between these values.

$$MSE = \frac{1}{n}\sum\nolimits_{t=1}^{n}\left(y_{real}(t) - y_{predicted}(t)\right)^2 \quad (2)$$

The mean absolute errors (MAE) is a properly-set up performance metric for forecasting troubles and is the most generally used accuracy degree in load forecasting. Utilising MAE as each a loss feature during training and a metric for overall performance assessment can be high-quality, as it aligns the goals of training and performance evaluation. However, this approach may lead to forecasts that are biased, since MAE does not directly address forecast bias.

$$MAE = \frac{1}{n}\sum\nolimits_{t=1}^{n}\left|y_{real}(t) - y_{predicted}(t)\right| \quad (3)$$

The subsequent criterion, RMSE is the standard deviation of the prediction errors (residuals). The RMSE is a measure of the dispersion of these residuals. The formula for calculating RMSE error:

$$RMSE = \sqrt{\frac{1}{n}\sum\nolimits_{t=1}^{n}\left(y_{real}(t) - y_{predicted}(t)\right)^2} \quad (4)$$

The R-squared (R^2) or coefficient of determination, is a metric that measures how well a regression model fits real data. It tells us how much of the variance in the dependent variable (actual value) the model explains.

$$R^2 = 1 - \frac{\sum_{t=1}^{n}\left(y_{real}(t) - y_{predicted}(t)\right)^2}{\sum_{t=1}^{n}\left(y_{real}(t) - y_{meanof\,thedataset}\right)^2} \quad (5)$$

3.3 Scenario

The test is designed to assess the overall performance of the N-BEATS version in training eventualities: centralized and distributed. Greenhouse temperature data is collected from various sources, then preprocessed to ensure consistency before training.

- Centralized training scenario:
  All data is centralized on a central server. The N-BEATS model is trained using the Adam optimization method with a fixed learning rate (0.001). The training process takes place in 50 epochs, with a fixed batch size. Evaluation metrics include MSE, RMSE, MAE, and R^2.
- Federated Learning (FL) training scenario:
  Data is distributed to independent greenhouses. Using the Federated Averaging algorithm to update global weights after each iteration. The number of participating greenhouses is 10, with data being divided unevenly among the devices. Each device performs local training for 5 epochs before sending the weights to the central server. Model synthesis takes place in 50 iterations. Evaluation metrics include MSE, RMSE, MAE, and R^2.

3.4 Result

The N-BEATS model trained by the centralized method achieves good results as the error metrics decrease with the number of epochs. Specifically, the R^2 value increases from −7.2988 to 0.7000, indicating that the model significantly improves its ability to explain the variance of the data (Table 1 and Fig. 3).

Table 1. Performance Metrics of Centralized vs. Federated Learning for Greenhouse Temperature Prediction

	MSE	RMSE	MAE	R^2
Centralized	0.0057	0.0757	0.0584	0.7
Federated Learning	0.0053	0.0975	0.0706	0.4561

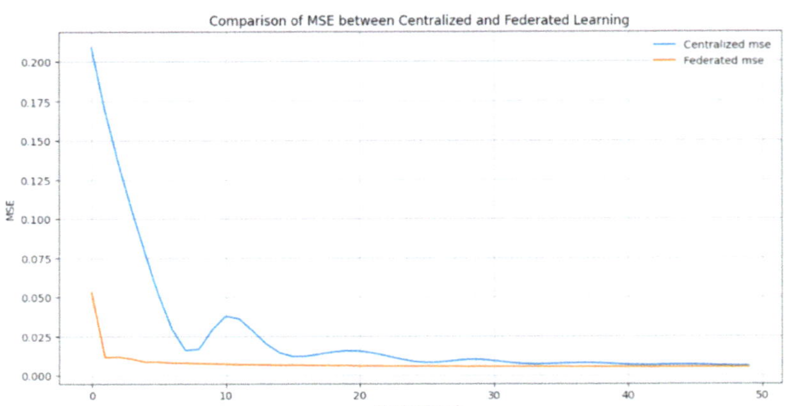

Fig. 3. Comparison of MSE between Centralized and Federated Learning

In the Federated Learning (FL) scenario, the model performance gradually improves with the number of training iterations. Initially, the loss (MSE) is quite high (0.0356 in round 1), but after 50 rounds, this value drops to 0.0053. Other error metrics also show a similar downward trend: RMSE drops from 0.1717 to 0.0975 but is still higher than that of centralized training (0.0757); MAE drops from 0.1292 to 0.0706 but is still larger than the centralized model (0.0584). Meanwhile, R^2 increases from −0.6881 to 0.4561, reflecting an improvement in the model's variance explanation, but still falls short of the accuracy of the centralized model.

Experimental results show that centralized training provides better predictive performance than Federated Learning. However, Federated Learning still shows significant potential when applied in distributed systems, especially in contexts with high requirements for data privacy and security. To improve the performance of Federated Learning, several approaches can be considered:

- Improve the model synthesis algorithm
- Adjust the weights of devices based on the importance of the data.
- Increasing the number of training cycles ensures better model convergence (Fig. 4).

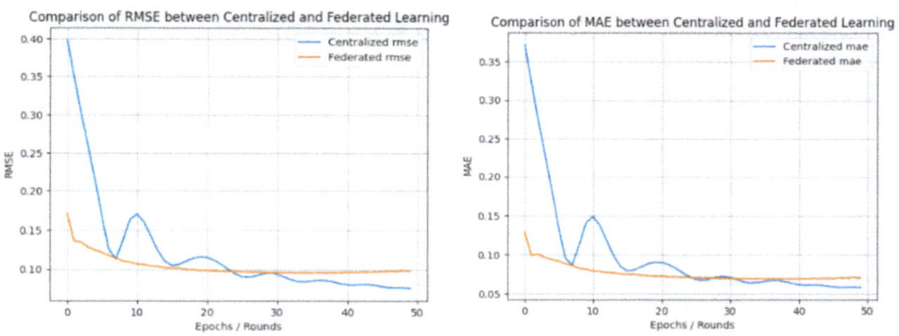

Fig. 4. Comparison of metrics between Centralized and Federated Learning

4 Conclusion

This study demonstrated the feasibility of combining N-BEATS with Federated Learning (FL) in the greenhouse temperature forecasting problem. Experimental results show that FL can achieve forecasting accuracy close to that of the centralized model, while ensuring data privacy and security by eliminating the need for centralized data sharing. However, there is still a performance gap between FL and centralized training methods, requiring improvements in model synthesis algorithms and training strategies to optimize performance in distributed environments.

In addition, this study also confirmed the ability of N-BEATS in time series forecasting, regardless of centralized or distributed training methods. Distributed training does not significantly degrade the model accuracy, paving the way for the application of machine learning techniques in decentralized data environments.

In the future, the research can be extended in several directions that can be applied to improve the performance of FL, including: Improving the model synthesis algorithm; Adjusting the weights of the devices according to the importance of the data; Increasing the number of training rounds to ensure better convergence.

References

1. Morales-García, J., Terroso-Sáenz, F., Cecilia, J.M.: A multi-model deep learning approach to address prediction imbalances in smart greenhouses. Comput. Electron. Agric. **216**, 108537 (2024)
2. Altikat, S.: Prediction of CO_2 emission from greenhouse to atmosphere with artificial neural networks and deep learning neural networks. Int. J. Environ. Sci. Technol. **18**(10), 3169–3178 (2021)
3. Esparza-Gómez, J., Guerrero-Osuna, H., Ornelas-Vargas, G., Luque-Vega, L.: Deep Learning for greenhouse internal temperature forecast. Int. J. Comb. Optim. Probl. Inform. **14**, 86–94 (2023)
4. Venkateswaran, D., Cho, Y.: Efficient solar power generation forecasting for greenhouses: a hybrid deep learning approach. Alex. Eng. J. **91**, 222–236 (2024)
5. Cletus, F., Anagu, E.J.: Comparative analysis of machine learning models for greenhouse microclimate prediction. Brill. Res. Artif. Intell. **4**, 162–175 (2024)
6. Douswekreo, S., Ndoumbe, J., Baba, A., Fedotova, M., Offole, F., Essola, D.: Solar irradiation and temperature prediction using LSTM neural network and solar energy potential in Northern Cameroon for solar electric vehicle application. J. Eng. **2024**, 1536889 (2024)
7. Eraliev, O., Lee, C.H.: Performance analysis of time series deep learning models for climate prediction in indoor hydroponic greenhouses at different time intervals. Plants **12**(12), 2316 (2023)
8. Oreshkin, B.N., Carpov, D., Chapados, N., Bengio, Y.: N-BEATS: neural basis expansion analysis for interpretable time series forecasting. In: International Conference on Learning Representations (2020)
9. Hemming, S., de Zwart, F., Elings, A., Petropoulou, A., Righini, I.: Autonomous Greenhouse Challenge, 2nd edn. Wageningen University & Research, 20 August 2020 (2019). https://doi.org/10.4121/uuid:88d22c60-21b3-4ea8-90db-20249a5be2a7

Rumor Detection Method Based on Multi-modal Mixture of Experts and Cross-modal Enhancement

Jianyong Yu[1,2], Xiuyu Li[1,2](✉), and Xue Han[3]

[1] Sanya Institute of Hunan University of Science and Technology, Sanya, China
yujyong@hnust.edu.cn, lixiuyu@mail.hnust.edu.cn

[2] School of Computer Science and Engineering, Hunan University of Science and Technology, Xiangtan, China

[3] Hunan Software Vocational and Technical University, Xiangtan, China

Abstract. Rumor spreaders are increasingly using multimedia content on social platforms to attract and mislead the public. To alleviate this issue, research on rumor detection has received extensive attention in recent years. Although current multimodal rumor detection models utilize both textual and visual features of rumors for detection, they overlook the problem of text-image mismatch in rumors and the differences in the extraction processes between textual data and image data (such as temporal information and spatial structure). This paper proposes a new rumor detection method based on Multi-modal Mixture of Experts with Cross-modal Enhancement (MMoE). This method uses a contrastive language-image pre-trained model to calculate the similarity between text and images. Through a gating unit, different experts are selected based on the similarity to ensure that the model's predictions are not misled by incorrect information. In addition, a visual encoder-decoder is employed to convert post images into descriptive text for data augmentation, leveraging additional text data to optimize the model's performance. Experimental results show that the rumor detection results of MMoE outperform six baselines and achieve an accuracy rate of 92.49% on a Chinese dataset.

Keywords: Rumor Detection · Multi-modal · Social Media · Deep Learning

1 Introduction

The vigorous development of the Internet and the continuous emergence of multimedia technologies have greatly promoted knowledge sharing, cultural integration, and economic development among different countries. The dissemination of real-time news and major messages has also broken through the limitations of distance and culture. Meanwhile, as a type of message, rumors have become more influential with the acceleration of their spread. As the saying goes, "It takes only one mouth to spread a rumor, but running around to refute it is exhausting." The rapid spread of rumors can have serious negative impacts, ranging from an individual's reputation to social order and even national security [1].

J. S. Park et al. (Eds.): FutureTech 2025, LNEE 1483, pp. 111–121, 2025.
https://doi.org/10.1007/978-981-95-1999-6_15

Traditional rumor detection methods mainly rely on salient features in textual data for classification. Some use traditional machine learning models, such as decision trees, while others use neural network-based models, such as Recurrent Neural Networks (RNNs) and Convolutional Neural Networks (CNNs). In addition, there are knowledge graphs [2], among others. With the advancement of multimedia technology and information dissemination technology, visual content has become an important component in attracting and misleading consumers due to its more credible narrative and faster spread [3].

Current multimodal rumor detection methods tend to abstract data within each modality using their respective feature extraction models, that is, converting data into feature vectors through methods such as neural networks, and then performing feature fusion. Although this approach can improve model accuracy, it overlooks the differences in heterogeneous data and feature extraction patterns of different models, as well as the mismatch between text and images in posts. To address these issues, this paper proposes a new rumor detection method based on multimodal mixture of experts and Cross-modal Enhancement. Considering the feature differences between text and images, and the fact that rumor detection is primarily text-based, this paper uses a visual encoder-decoder model [4] to convert images from the same post into descriptive text to extract textual features for data augmentation. Additionally, the CLIP (Contrastive Language-Image Pre-training) model is used to project text and images into the same feature space and calculate their similarity, providing guidance for the gating network to select appropriate expert networks. Moreover, the expert networks are transformed into masked expert networks to ensure that their inputs correspond to the correct modality information. Experimental results show that the multimodal data conversion network proposed in this paper outperforms recently proposed methods and baseline methods on both Chinese and English datasets.

2 Related Work

Regarding the definition of rumors, researchers have provided a relatively consistent definition through decades of social science investigations: "A rumor is information whose truthfulness has not yet been verified at the time of dissemination" [5].

Lin et al. [6] proposed a zero-shot framework based on prompt learning, representing rumors spreading on social media as independent propagation threads and modeling domain-invariant structural features from them to incorporate into the representation of influential community structures. Then, they designed a hierarchical prompt encoding to learn the semantic features of prompts and rumor data. Additionally, they used a novel virtual response enhancement method to improve model training. Yuan et al. [7] proposed a global-local attention network that first fuses the original post and reposts to obtain a better semantic representation and then uses the heterogeneous graph structure among the original post, reposts, and users for rumor detection. Zheng et al. [8] was the first to propose a framework that combines textual features, visual features, and graph structure features and adopts co-attention for fusion. It also introduces latent relationship completion in social networks, multimodal alignment, and adversarial learning, achieving excellent results in rumor dete ction. Li et al. [9] introduced a focal reasoning model with two main modules: coarse-grained selection and fine-grained reasoning. In summary, it first selects sentence-level features that are focused on by users and then uses a relational attention module to construct a fully connected graph between sentences and focal points to capture fine-grained relationships and perform reasoning.

With the development of natural language processing and the continuous emergence and improvement of large language models (LLMs), researchers have begun to apply LLMs to multimodal rumor detection. Yang et al. [10] proposed a novel LLM-enhanced multiple instance learning method. This model uses stance explanations generated by LLMs to strengthen the interaction representations between topics, posts, and any two elements during information propagation. It also designs a hierarchical stance tree attention mechanism to leverage LLM-generated stance explanations to improve rumor detection in binary classifiers.

3 Methodology

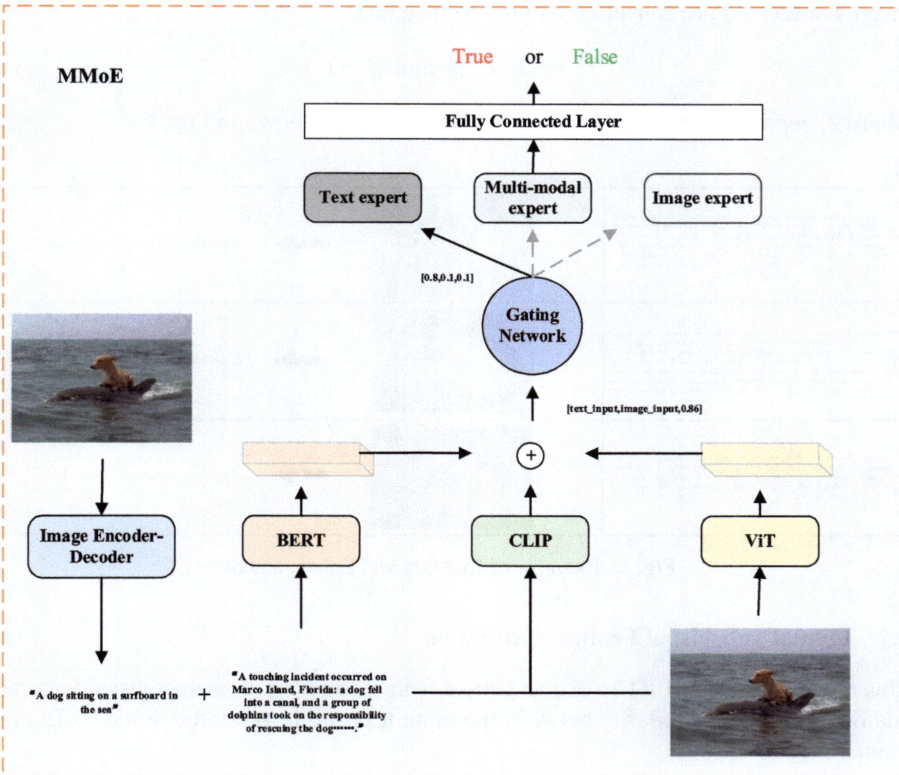

Fig. 1. The framework of the proposed MMoE

The method in this paper consists of four parts, and its structure (see Fig. 1). First, a visual encoder-decoder is used to convert images into descriptive text. Then, BERT and ViT are utilized to extract textual features and visual features respectively. The similarity between the input text and image is determined through the CLIP pre-trained model. Subsequently, the mixed data and similarity are input into the gating network to obtain the output weights of each expert. Next, the outputs of the top K experts with the highest weights are selected and weighted summed. Finally, the result is input into a fully connected layer for classification. A mask matrix is added to the expert network to ensure that it receives the corresponding modal input.

3.1 Cross-modal Enhancement

This paper uses image captions [11, 12] to convert images into text. Image captioning belongs to a type of visual encoder-decoder model. This model uses a Transformer-based pre-trained visual model as the encoder (e.g., ViT [13], Swin [14]) and a pre-trained language model as the decoder (e.g., GPT, BERT). First, the image is converted into image features $H_I = [i_1, ..., i_l]$ through the encoder using image captioning. Then, the image features H_I are converted into descriptive text.

$$T_I = Decoder(Encoder(I)) \tag{1}$$

where T_I represents the descriptive text of the image, as shown in Fig. 2.

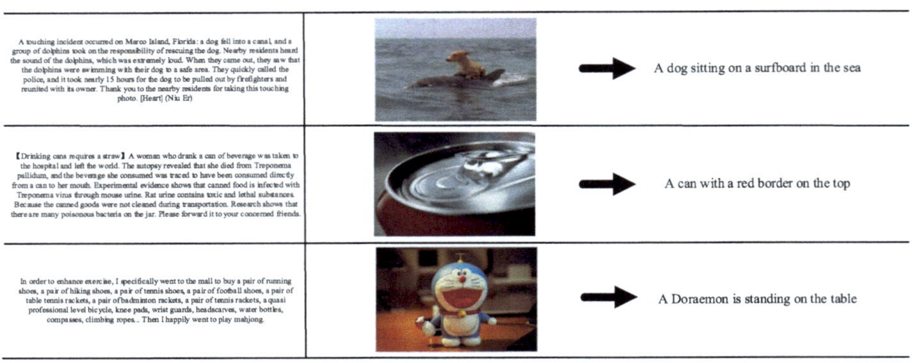

Fig. 2. Example of Cross-modal Enhancement

3.2 Textual and Visual Feature Extraction

This module utilizes BERT to extract textual features and ViT to extract visual features, and determines the similarity between the input text and image through the CLIP pre-trained model.

$$H_T = BERT(T, T_I) \tag{2}$$

$$H_I = ViT(I) \tag{3}$$

Here, $H_T = [t_1, ..., t_l]$, where t_i represents the feature of the i-th word, l represents the text length, and the dimensionality of t_i is $d = 768$. In order for the gating network to accurately determine the degree of text-image mismatch, the CLIP model is introduced to calculate the similarity between the image and the text.

$$S = CLIP(T, I) \tag{4}$$

where S is a constant representing the similarity between the text and the image of the post.

3.3 Multimodal Mixture of Experts

The Mixture of Experts (MoE) is an ensemble learning model based on the "divide-and-conquer strategy". The process of this model is to decompose a complex task into multiple subtasks, with each subtask being handled by an "expert" model, and input samples are dynamically allocated to the most suitable expert through a gating network. The multimodal mixture of experts network in this module improves upon the basic mixture of experts network by adding different masking layers to each expert network to control the data of the corresponding modality accepted by the expert network.

The main purpose of this module is to assign input features to multiple expert networks through a dynamic routing mechanism and sum the outputs of each expert with weights. First, the text features, visual features, and similarity scores are simply concatenated and input into the gating network.

$$x = concat(H_T, H_I, S) \tag{5}$$

Then, the gating network generates weights for each expert and ensures that the sum of all weights is 1. Finally, the mixed features are input into the expert network. The formula is as follows:

$$H' = softmax\left(\sum_{i=1}^{N} G_{t,i}(x) \cdot E_i(MASK(x))\right) \tag{6}$$

where $x \in \mathbb{R}^d$ represents an input vector with a feature dimension of d. $MASK(x)$ is a masking matrix that ensures each expert receives the corresponding modality information. $G_{t,i}(x)$ is used by the gating network for weight allocation, and $E_i(x)$ is the output of the i-th expert. Each expert is an independent feedforward neural network (FFN) with a masking matrix, receiving the same input and having the same structure but different parameters.

3.4 Loss Function

The features H' output by the multimodal mixture of experts network are input into a fully connected layer to predict whether the post is a rumor.

$$\hat{y}_i = Sigmoid(FC(H')) \tag{7}$$

where $\hat{y}_i$ is the final prediction output of the model, $FC(\cdot)$ represents the fully connected layer, and $Sigmoid(\cdot)$. . is the activation function, which can project the output of the fully connected layer between [0,1] as the predicted probability. Finally, the binary cross-entropy loss between the predicted label and the true label is calculated.

$$L(\theta) = -(y \log(1 - \hat{y}_i) + (1 - y) \log \hat{y}_i) \tag{8}$$

4 Experiments

4.1 Dataset

This paper compares MMoE with other methods on three large datasets: PHEME [15], Weibo [16], and Weibo-MM [17]. The Chinese dataset Weibo is from the Weibo platform. PHEME is from the Twitter platform and contains five news events. Each piece of data includes text, the corresponding pictures of the text, and user information. Table 1 shows the dataset information.

Table 1. Statistical information of the dataset

Dataset	Non-Rumors	Rumors	Images	Comments
PHEME	1428	590	2018	7388
Weibo	877	590	1467	4534
Weibo-MM	4640	4488	9128	9128

4.2 Parameter Settings

This paper divides each of the three datasets into training sets, validation sets, and test sets in a ratio of 7:2:1. For the preprocessing of text in the datasets, through statistics, it is found that more than 95% of the text characters in the posts of the three datasets are less than 160, and they are concentrated between 100 and 160 characters. Therefore, the WordPiece algorithm is used to tokenize the datasets, and the maximum text length is set to 160. The features obtained by the text through BERT are 768-dimensional, and the visual features are also initialized to be 768-dimensional, which can accommodate more information while not making the number of model parameters too large. Since it is a binary classification task, the Sigmoid function is used as the activation function, BCELoss is used as the loss function, and the optimizer AdamW is selected. The learning rate is set to 1e-5, the batch size is set to 12, and the training is carried out for 30 epochs. The commonly used accuracy, precision, recall, and F1 Score are used as evaluation metrics.

4.3 Experimental Analysis

We compared our method with the following baselines:

MVAE [18]: MVAE employs a multimodal Variational Autoencoder (VAE) to extract features from different modalities and complete rumor detection by predicting the results through a binary classifier.

EANN [19]: EANN is a multimodal fake news detection model based on GAN (Generative Adversarial Networks). Ituses GAN to model the relationships between different events.

BERT [20]: BERT is a currently popular pre-trained model. Based on Transformer, it extracts the general semantic information of each word. Moreover, after MLM pre-training, it can be easily applied to various downstream tasks related to text.

MFAN [8]: MFAN first attempts to fuse text, visual, and graph features together. It improves social graph feature learning by enhancing the graph topology and neighborhood aggregation process through link prediction and also incorporates adversarial learning to enhance the model's performance.

MRML [21]: MRML adopts metric-based triplet learning to extract intra-modal associations between rumors and non-rumors in each modality and utilizes contrastive learning to capture inter-modal relationships.

ELKP [22]: ELKP constructs a knowledge prompt composition module that can dynamically learn external knowledge related to promotion and selectively integrate external knowledge into prompts. In addition, this model performs joint multi-task collaborative optimization between the prediction objective of prompt learning and the classification objective of the discriminative model.

Table 2. Comparative experimental results

Dataset	Method	Accuracy	Precision	Recall	F1
PHEME	MVAE	77.62	73.49	72.25	72.77
	EANN	77.13	71.39	70.07	70.44
	BERT	79.64	75.32	71.45	71.40
	MFAN	88.73	87.07	85.61	86.16
	MRML	84.26	79.34	77.95	78.64
	ELKP	89.22	88.41	87.69	88.04
	MMoE	**90.07**	**94.42**	**91.04**	**92.70**
Weibo	MVAE	71.67	70.52	70.12	70.34
	EANN	80.96	80.19	79.68	79.84
	BERT	82.45	82.16	89.48	80.36
	MFAN	88.95	88.91	88.13	88.33
	MRML	89.70	89.84	88.76	89.30
	ELKP	90.25	89.33	90.56	89.94
	MMoE	**92.49**	**93.12**	**95.14**	**94.18**

(continued)

Table 2. (*continued*)

Dataset	Method	Accuracy	Precision	Recall	F1
Weibo-MM	MVAE	74.50	80.13	71.94	75.81
	EANN	78.29	82.74	69.71	75.67
	BERT	82.41	85.45	76.90	80.95
	MFAN	89.22	88.75	85.13	86.90
	MRML	85.24	87.21	81.08	84.03
	ELKP	89.68	88.70	87.92	88.30
	MMoE	**91.81**	**92.59**	**94.59**	**93.58**

From the experimental results (see Table 2), it can be seen that on these three datasets, the performance of the MMoE model is superior to that of other models. The performance of BERT is significantly better than that of MVAE and EANN, demonstrating the excellent generalization ability of pre-trained language models, reflecting the impact of data volume on rumor detection. Compared with BERT, MRML reflects the advantages of contrastive learning in rumor monitoring. In addition, MFAN outperforms previous models, indicating that considering the propagation structure and modal alignment is beneficial for rumor detection. ELKP outperforming MFAN indicates the effectiveness of professional domain knowledge in rumor detection and that a modular structure is conducive to improving the generalization performance of the model. The MMoE in this paper outperforms ELKP. From the perspective of training content, BERT has obtained a lot of general knowledge during pre-training, providing guidance for the model's rumor detection. The model used by the visual encoder-decoder has completed the generation training of text-image pairs on a large amount of data, which is equivalent to providing additional dataset support for the model. CLIP provides the basis for the model to determine the mismatch between images and texts, enabling the multimodal mixture of experts network to make accurate predictions on multimodal data and improving the model's rumor detection ability.

4.4 Ablation Analysis

To verify the effectiveness of different modules in MMoE we designed several submodels. The experimental results are shown in Table 3:

w/o BERT: Replace the pre-trained BERT with Word2Vec to extract text features on the basis of MMoE.

w/o ViT: Replace the pre-trained ViT model with ResNet50 to extract visual features on the basis of MMoE.

w/o CLIP: Remove the CLIP model on the basis of MMoE and only use text features and visual features.

w/o Gate: Remove the gating network on the basis of MMoE, simply concatenate the text features, visual features and similarity scores and directly input them into the mixture of experts module. At the same time, set the weight of each expert network to 0.3.

Table 3. Ablation Study results

	PHEME		Weibo		Weibo-MM	
Method	Accuracy	F1 Score	Accuracy	F1 Score	Accuracy	F1 Score
w/o BERT	84.62	83.23	85.53	87.76	86.61	88.76
w/o ViT	**91.25**	89.92	91.54	90.04	87.46	90.15
w/o CLIP	87.62	85.46	88.19	87.51	90.84	89.93
w/o Gate	84.54	85.13	86.33	88.27	88.21	87.94
MMoE	90.07	**92.70**	**92.49**	**94.18**	**94.59**	**93.58**

From the experiments in the table, it can be seen that when using the traditional Word2Vec word embedding model to extract features from text, the accuracy rates on the three datasets reached 84.62%, 85.53%, and 86.61% respectively. After using the pre-trained BERT model, the accuracy rates increased to 90.07%, 92.49%, and 94.59% respectively, which were 5.45%, 6.96%, and 7.98% higher. This means that BERT can better capture the deep features of text. When replacing the ViT pre-trained model with ResNet50 to extract visual features, although the F1 score decreased by 2.78% on the PHEME dataset compared to the original architecture, the accuracy rate increased by 1.18%. This indicates that the ViT pre-trained model and the ResNet50 residual connection network each have their own advantages and disadvantages. After removing the CLIP model, the gating network will not be able to obtain accurate image-text similarity information. Without the similarity score provided by the CLIP model, the performance of the model declines. After removing the gating network, the accuracy and F1 score of the model decreased on all three datasets. This shows that simply concatenating different modal features and inputting them into different parameter networks for "voting"-style integration cannot improve the performance of the model; instead, it will reduce the model's performance.

5 Conclusion

This paper proposes a model based on Multimodal Mixture of Experts and Cross-modal Enhancement (MMoE) for rumor detection, which combines text and images. To address the differences in text and image feature extraction among different models, a visual encoder-decoder is introduced to convert images into text, providing effective additional data for the model. To solve the problem of text-image mismatch in posts, CLIP is incorporated to calculate similarity, and a masked mixture of experts network is designed to ensure that the model makes predictions based on valid information. The experimental results on three datasets confirm the advantages of the multimodal mixture of experts model in this paper. In future work, it is planned to optimize the visual encoder-decoder model to obtain more accurate and rich descriptive text from images. At the same time, considering the development trend of LLMs, integrating them into the rumor detection framework as an external knowledge base or for text-image conversion may also achieve good results.

Acknowledgment. This work is supported by The National Social Science Fund of China under Grant No. 20BXW096.

References

1. Abdali, S., Shaham, S., Krishnamachari, B.: Multi-modal misinformation detection: approaches, challenges and opportunities. ACM Comput. Surv. **57**(3), 1–29 (2024)
2. Sun, M., Zhang, X., Ma, J., Xie, S., Liu, Y., Philip, S.Y.: Inconsistent matters: a knowledge-guided dual-consistency network for multi-modal rumor detection. IEEE Trans. Knowl. Data Eng. **35**(12), 12736–12749 (2023)
3. Qi, P., Cao, J., Yang, T., Guo, J., Li, J.: Exploiting multi-domain visual information for fake news detection. Paper Presented at the 2019 IEEE International Conference on Data Mining (ICDM) (2019)
4. Kumar, A.: The Illustrated Image Captioning Using Transformers. Ankur-NLP Enthusiast (2022)
5. Zubiaga, A., Aker, A., Bontcheva, K., Liakata, M., Procter, R.: Detection and resolution of rumours in social media: a survey. ACM Comput. Surv. **51**(2), 1–36 (2018)
6. Lin, H., et al.: Zero-shot rumor detection with propagation structure via prompt learning. Paper Presented at the AAAI (2023)
7. Yuan, C., Ma, Q., Zhou, W., Han, J., Hu, S.: Jointly embedding the local and global relations of heterogeneous graph for rumor detection. Paper Presented at the 2019 IEEE International Conference on Data Mining (ICDM) (2019)
8. Zheng, J., Zhang, X., Guo, S., Wang, Q., Zang, W., Zhang, Y.: MFAN: multi-modal feature-enhanced attention networks for rumor detection. Paper Presented at the IJCAI (2022)
9. Li, J., et al.: Focusing on relevant responses for multi-modal rumor detection. IEEE Trans. Knowl. Data Eng. **36**(11), 6225–6236 (2024)
10. Yang, R., Ma, J., Gao, W., Lin, H.: LLM-enhanced multiple instance learning for joint rumor and stance detection with social context information. ACM Trans. Intell. Syst. Technol. **16**(3), 1–27 (2025)
11. Sasibhooshan, R., Kumaraswamy, S., Sasidharan, S.: Image caption generation using visual attention prediction and contextual spatial relation extraction. J. Big Data **10**(1), 18 (2023)
12. Xu, L., Tang, Q., Lv, J., Zheng, B., Zeng, X., Li, W.: Deep image captioning: a review of methods, trends and future challenges. Neurocomputing **546**, 126287 (2023)
13. Dosovitskiy, A., et al.: An Image is Worth 16x16 Words: Transformers for Image Recognition at Scale. Computing Research Repository (2021)
14. Liu, Z., et al.: Swin transformer: hierarchical vision transformer using shifted windows. Paper Presented at the Proceedings of the IEEE/CVF International Conference on Computer Vision (2021)
15. Zubiaga, A., Liakata, M., Procter, R.: Exploiting context for rumour detection in social media. Paper Presented at the Social Informatics: 9th International Conference, SocInfo 2017, Oxford, UK, 13–15 September 2017, Proceedings, Part I 9 (2017)
16. Song, C., Yang, C., Chen, H., Tu, C., Liu, Z., Sun, M.: CED: credible early detection of social media rumors. IEEE Trans. Knowl. Data Eng. **33**(8), 3035–3047 (2019)
17. Jin, Z., Cao, J., Guo, H., Zhang, Y., Luo, J.: Multimodal fusion with recurrent neural networks for rumor detection on microblogs. In: ACM International Conference on Multimedia (2017)
18. Khattar, D., Goud, J.S., Gupta, M., Varma, V.: MVAE: multimodal variational autoencoder for fake news detection. Paper Presented at the The World Wide Web Conference (2019)

19. Wang, Y., Ma, F., Jin, Z., Yuan, Y., Jha, K.: EANN: event adversarial neural networks for multi-modal fake news detection. Paper Presented at the ACM SIGKDD International Conference (2018)
20. Devlin, J.: BERT: pre-training of deep bidirectional transformers for language understanding. arXiv preprint arXiv:1810.04805 (2018)
21. Peng, L., Jian, S., Li, D., Shen, S.: MRML: multimodal rumor detection by deep metric learning. In: IEEE International Conference on Acoustics, Speech, Signal Processing, pp. 1–5 (2023)
22. Yan, Y., Zheng, P., Wang, Y.: Enhancing large language model capabilities for rumor detection with knowledge-powered prompting. Eng. Appl. Artif. Intell. **133**, 108259 (2024)

Soft Porn Identification Based on Social Media-- User Engagement Perspective

Jingmeng Dou, Xuepeng Xu, and Xiaofeng Yu(✉)

School of Business, Nanjing University, Nanjing, China
xiaofengyu@nju.edu.cn

Abstract. Pornography has existed since the early days of the Internet. With the rapid growth of video-sharing platforms, a new form of pornography, online soft pornography, has emerged. While its harms are evident, few studies have focused on its conceptualization or identification. This paper proposes a deep learning method to identify soft pornography using a new perspective—user engagement. We utilize video data from Bilibili's dance section to measure user engagement from both content creators and users based on interactive characteristics and build a learning model. The results show that the model based on user engagement outperforms the text-based model (87.6% vs. 83.5%) in identifying soft pornography, and the fusion model further improves the performance (88.5%). Principal component analysis reveals that in selective attention-driven participation, users of soft pornography exhibit a significantly higher interest index. This study provides a detailed understanding of user engagement with soft pornography on video platforms and offers new insights for content regulation.

Keywords: online soft pornography · social media · user engagement

1 Introduction

Pornography has been intertwined with the internet since its inception, with its online presence expanding rapidly. The rise of mobile devices and the internet's accessibility, affordability, and anonymity [1] have further facilitated its consumption. Social media has amplified this trend, enabling interactive engagement with suggestive content.

The prevalence of soft pornography on social media poses risks, highlighting the need for a method to detect soft pornographic content on social platforms. Current detection methods fall into three categories. Survey studies on online pornography behavior explore psychological traits and influencing factors through surveys or experiments [2], but suffer from biased samples and self-reporting biases. Deep learning-based detection applies techniques to detect pornographic text and videos, but is less effective for subtle, lower-density content on social media [3]. User behavior-based detection analyzes patterns when visiting pornographic websites [4], but is not directly applicable to the diverse content on social media.

This study proposes a novel approach to identifying soft pornography by analyzing user engagement patterns. Prior research suggests that users exhibit distinct behaviors

J. S. Park et al. (Eds.): FutureTech 2025, LNEE 1483, pp. 122–129, 2025.
https://doi.org/10.1007/978-981-95-1999-6_16

when interacting with pornographic content [4], often displaying specific emotional and attentional biases. User engagement, defined as the degree of interaction with digital content [5], has been extensively studied in various online contexts [6]. By leveraging engagement metrics, we aim to develop a classification model for detecting soft pornography on social media.

Focusing on Bilibili, we analyze 9,976 dance videos (2020–2021) and user engagement data. Our results show that engagement-based models outperform text-based models in detecting soft pornography. Additionally, distinct behavioral patterns emerge across non-engagement, selective engagement, and sustained engagement, revealing the differences in user behavior patterns when interacting with different videos.

2 Theoretical Foundation

2.1 Online Pornography on Social Media

The rise of mobile internet and social media has reshaped online pornography, making it more accessible and harder to regulate [3]. Unlike traditional adult content, social media-based "soft pornography" stays within legal gray areas, leveraging suggestive imagery to attract attention and generate engagement. With low entry barriers and widespread social media use, its reach far exceeds that of traditional adult websites.

Soft pornography spreads through clickbait, targeted recommendations, and monetized platforms. Its risks include addiction, financial exploitation, and potential links to online sexual crimes [7]. Studies show that problematic pornography use affects a significant portion of users, with 7.6% reporting compulsive behaviors [2]. While regulations exist, enforcement remains challenging due to the subtle nature of soft pornography and platforms' reliance on engagement-driven algorithms. Existing detection methods struggle with implicit content, highlighting the need for improved moderation strategies [3].

2.2 User Engagement Theory

User engagement is a type of experience influenced by factors such as attention, emotional arousal, interactivity, and user control [8]. It involves perception (attention level), cognition (motivation and preference), experience (immersion), and action (interaction or content consumption) [9].

O'Brien and Toms [8] define four engagement stages: initiation (triggered by emotional resonance), engagement (sustained by attention and interaction), disengagement (due to loss of interest or external interruptions), and re-engagement (when content re-attracts the user). In the fast-paced social media environment, engagement is often described as discovery, interest, usage, and interaction [9]. Engagement is typically measured through attention, emotional involvement, persistence, and novelty [10]. It can be assessed using subjective indicators (self-reported opinions) and objective behavioral metrics (e.g., clicks) [11].

3 Research Design and Methodology

This research aims to create a model to identify soft pornography content in Bilibili Dance Section videos. It first collects and annotates video data, then builds different classification models based on user engagement data and video titles. These models are combined into a fused model to improve accuracy, and their performance is compared to find the most effective approach.

We collected dance videos from Bilibili (B站) from 2020 to 2021, chosen for its regulatory scrutiny in 2020 and its comprehensive user behavior data.

User engagement features were extracted from video metadata, and videos were manually annotated. These features served as independent variables, while annotation results were dependent variables. A deep learning model using video-level features was designed to classify implicit adult content.

3.1 Data Collection and Video Annotation

A total of 10,000 dance videos (2020–2021) were collected, with 9,976 valid entries after removals. Sorting by publication date reduced sample bias. Metadata included video details (URLs, titles, tags, etc.) and engagement metrics (views, likes, shares, etc.). Videos were manually annotated for implicit adult content using a standardized manual [12]. Three trained annotators achieved a reliability score of 0.956.

In recent years, researchers have proposed a methods for measuring collective user engagement features by leveraging more comprehensive social media data [13]. Based on the user engagement measurement factors summarized by O'Brien and Toms [8], they identified data available on social media to quantify these factors and demonstrated that this approach aligns with the theoretical framework of user engagement [6]. In this study, users are categorized into content creators and content consumers. Individual engagement features for content creators and collective engagement features for content consumers are constructed using metadata from the collected videos.

3.2 Measurement of Content Creator Engagement Features

In constructing the content creator engagement feature metrics, this study focuses on three core categories: motivation, emotion, and interaction, to more precisely identify and analyze the creation and dissemination patterns of implicit adult content. Video titles often contain sexually suggestive elements to attract user engagement and traffic. Creators use attention-grabbing titles, tags, and descriptions to boost views. This study measures the creator's motivation using special symbols in titles (e.g., ♥, ★, ♂, ♀), number of tags, and description length.

Additionally, based on previous studies, interactivity is an important indicator of engagement (Liu et al., 2019). Therefore, the number of second-person pronouns in video titles is used to assess the creator's willingness to interact. Tables 1 and 2 show the engagement metrics and descriptive statistics for creators of implicit adult content and regular videos.

Social platforms encourage feedback through comments and likes [14]. Content with sexual appeal receives higher engagement, including more likes and views [12].

Table 1. Descriptive Statistics of Soft Pornography Content Creators' Engagement Features

Feature	Measurement	Min	Median	Mean	Max	Standard Deviation
Motivation	Number of special symbols in title	0	5	5.4134	43	3.3426
	Number of video tags	0	9	9.7344	21	4.1085
	Number of words in video description	1	53	67.874	261	60.6817
Emotion	Sentiment polarity of video title	0	0.8216	0.6985	1	0.3076
Interaction	Number of second-person pronouns in title	0	0	0.2678	3	0.481

Table 2. Descriptive Statistics of Regular Content Creators' Engagement Features

Feature	Measurement	Min	Median	Mean	Max	Standard Deviation
Motivation	Number of special symbols in title	0	5	5.2533	49	3.6114
	Number of video tags	1	9	9.817	23	4.0958
	Number of words in video description	1	48	69.4425	269	70.5648
Emotion	Sentiment polarity of video title	0.0001	0.8569	0.7186	1	0.3121
Interaction	Number of second-person pronouns in title	0	0	0.1486	4	0.3819

Therefore, feedback is measured by video views, likes, and Bilibili's "coin" system, which indicates video popularity.

Interaction refers to the relationship between users or between users and creators, often through shares and comments. On Bilibili, bullet comments serve as key real-time interaction metrics. This study uses the number of favorites to measure user interest. Compulsive use of online pornography increases the likelihood of re-engagement, and attention bias makes it harder for users to disengage [15]. On Bilibili, favorites reflect users' re-engagement potential. Tables 3 and 4 show the engagement feature indicators and descriptive statistics for soft pornography and regular video content users.

Table 3. Descriptive Statistics of Soft Pornography Video Content Users' Engagement Features

Feature	Measurement	Min	Median	Mean	Max	Standard Deviation
Feedback	Video Views	115,000	289,000	444247.7	12057000	545756.1
	Video Likes	194	11,000	14,844.63	1,196,000	24,682.19
	Video Coins	11	2,472	4,266.89	196,000	7,345.12
Interaction	Video Shares	12	415	1,002.57	86,000	2,778.47
	Video Comments	7	366	540.6361	28,788	869.4696
	Video Bullet Comments	1	206	439.3467	76,000	1,447.31
Interest	Video Favorites	274	10,000	13,697.60	284,000	13,867.83

Table 4. Descriptive Statistics of General Video Content Users' Engagement Features

Feature	Measurement	Min	Median	Mean	Max	Standard Deviation
Feedback	Video Views	115,000	273,000	554664.2	23860000	889868.7
	Video Likes	65	13,000	28,660.31	2,081,000	60,005.38
	Video Coins	3	2,850.50	8,451.16	1,096,000	26,922.86
Interaction	Video Shares	6	928	3,034.31	331,000	9,986.82
	Video Comments	2	451	869.9179	113,095	2,291.42
	Video Bullet Comments	1	367	1,340.64	202,000	5,342.24
Interest	Video Favorites	18	4,751	9,737.00	615,000	19,587.78

4 Result and Analysis

Among 9,976 dance videos, 4,485 (44.96%) were labeled as soft pornography, and 5,491 (55.04%) as regular videos, with an inter-annotator reliability of 0.956. To evaluate performance, we compared text-based models (TF-IDF from titles) with user engagement-based models using Naïve Bayes, Linear Regression, Random Forest, and XGBoost. A 60/40 train-test split was applied. Table 5 show the results.

User engagement-based models consistently outperformed text-based ones, especially in predicting soft pornography videos. All four classifiers achieved higher accuracy, precision, recall, and F1-score, confirming the approach's effectiveness and feasibility.

Table 5. Classification Model Results Based on Difference Features

	User Engagement Features				Text Features			
Classification Model	Accuracy	Precision	Recall	F1-score	Accuracy	Precision	Recall	F1-score
Naive Bayes	0.737	0.920	0.651	0.762	0.853	0.789	0.871	0.828
Linear Regression	0.849	0.830	0.835	0.832	0.834	0.704	0.902	0.791
Random Forest	0.876	0.827	0.890	0.857	0.835	0.737	0.875	0.800
XGBoost	0.846	0.810	0.844	0.826	0.790	0.617	0.876	0.724

Next, this paper selects the Random Forest classifier, which has the highest classification accuracy, to compare the performance of text features, user engagement features, and a combined feature model that integrates both. The results are shown in Table 6. All evaluation metrics of the model based on the combined features have improved.

Table 6. Classification Model Results Based on Combined Features

Classification Model	Accuracy	Precision	Recall	F1-score
Text Features	0.835	0.737	0.875	0.800
User Engagement Features	0.876	0.827	0.890	0.857
Combined Features	0.885	0.841	0.896	0.867

Due to the high correlation between collective engagement measures of content users, Principal Component Analysis (PCA) was applied to reduce dimensions and identify engagement types. Table 7 show PCA results for soft pornography and regular videos. These dimensions can be defined as non-engagement, sustained attention-driven engagement, and selective attention-driven engagement [9, 13].

The first principal component represents sustained engagement, driven by interactions like comments, likes, and tipping. The second reflects selective engagement, measuring content alignment with user interests. The third, with negative loading on collective engagement, indicates a lack of interest or homepage-driven exposure. Results show minimal differences in sustained engagement between soft pornography and regular videos. However, soft pornography users exhibit higher selective engagement, suggesting a greater likelihood of re-engagement and compulsive use. They are also more likely to tip but less inclined to share or comment due to privacy concerns.

Table 7. Principal Component Analysis Results

		Soft Pornography Videos			Regular Videos		
	Measurement	Sustained	Selective	Non	Sustained	Selective	Non
Feedback	Video plays	0.421	0.033	− 0.236	0.420	0.461	− 0.119
	Video likes	0.431	− 0.158	0.252	0.412	0.255	− 0.116
	Video coin count	0.419	0.054	− 0.123	0.411	− 0.491	− 0.045
Interaction	Video share count	0.409	− 0.110	− 0.050	0.398	0.372	− 0.049
	Video comment count	0.347	− 0.236	0.699	0.257	− 0.054	− 0.421
	Video bullet comment count	0.151	0.950	0.248	0.307	0.019	0.886
Interest	Video collection count	0.389	0.039	− 0.559	0.408	− 0.582	− 0.070

5 Discussion and Conclusion

This study contributes to soft pornography detection on social media by applying user engagement theory. It demonstrates the effectiveness of engagement-based identification compared to text-based methods and highlights key behavioral differences between soft pornography and regular content. Notably, selective engagement—reflected in interest and interactive behaviors—emerges as a distinguishing factor.

A key contribution is the integration of both individual and collective engagement behaviors, providing a comprehensive view of social media participation. By considering engagement patterns from both content creators and users, the study offers insights into how personal actions influence group dynamics. However, limitations exist. The study focuses on identifying published content, limiting its ability to detect soft pornography before dissemination. Future research should explore predictive models for early detection and intervention.

References

1. de Alarcon, R., de la Iglesia, J.I., Casado, N.M., Montejo, A.L.: Online porn addiction: what we know and what we don't—a systematic review. J. Clin. Med. **8**(1) (2019). Art no. 91. https://doi.org/10.3390/jcm8010091
2. Ross, M.W., Mansson, S.-A., Daneback, K.: Prevalence, severity, and correlates of problematic sexual internet use in Swedish men and women. Arch. Sex. Behav. **41**(2), 459–466 (2012). https://doi.org/10.1007/s10508-011-9762-0

3. Kotta, I., Kalcza-Janosi, K., Marschalko, E.-E., Bandi, S., Bibok, B.: Online Sexual Activity Scale (OSAS): going beyond the nosological polemics of compulsivity vs addiction. Int. J. Ment. Heal. Addict. **21**(5), 3276–3292 (2023). https://doi.org/10.1007/s11469-022-00791-4
4. Jianxun, C., Yiqun, L.: Pornography web site identification based on user behavior analysis. J. Comput. Res. Dev. **50**, 430–436 (2013)
5. Laurel, B.: Computers as Theatre. Addison-Wesley, Boston (2013)
6. O'Brien, H.L., Toms, E.G.: The development and evaluation of a survey to measure user engagement. J. Am. Soc. Inform. Sci. Technol. **61**(1), 50–69 (2010). https://doi.org/10.1002/asi.21229
7. Gola, M., Lewczuk, K., Skorko, M.: What matters: quantity or quality of pornography use? Psychological and behavioral factors of seeking treatment for problematic pornography use. J. Sex. Med. **13**(5), 815–824 (2016)
8. O'Brien, H.L., Toms, E.G.: What is user engagement? A conceptual framework for defining user engagement with technology. J. Am. Soc. Inform. Sci. Technol. **59**(6), 938–955 (2008)
9. Peters, C., Castellano, G., De Freitas, S.: An exploration of user engagement in HCI. In: Proceedings of the International Workshop on Affective-Aware Virtual Agents and Social Robots, pp. 1–3 (2009)
10. Baldauf, D., Burgard, E., Wittmann, M.: Time perception as a workload measure in simulated car driving. Appl. Ergon. **40**(5), 929–935 (2009). https://doi.org/10.1016/j.apergo.2009.01.004
11. Attfield, S., Kazai, G., Lalmas, M., Piwowarski, B.: Towards a science of user engagement (position paper). In: WSDM Workshop on User Modelling for Web Applications, vol. 1. Citeseer (2011)
12. Yan, K., Salmon, J., Aubrey, J.S.: A sexy post a day brings the "Likes" your way: a content analytic investigation of sexualization in fraternity Instagram posts. Sex. Cult. Interdiscip. J. **26**(2), 685–706 (2022). https://doi.org/10.1007/s12119-021-09915-9
13. Liu, X., Zhang, B., Susarla, A., Padman, R.: Go to YouTube and call me in the morning: use of social media for chronic conditions. MIS Q. **44**(1), 257–283 (2019)
14. Trekels, J., Karsay, K., Eggermont, S., Vandenbosch, L.: How social and mass media relate to youth's self-sexualization: taking a cross-national perspective on rewarded appearance ideals. J. Youth Adolesc. **47**(7), 1440–1455 (2018). https://doi.org/10.1007/s10964-018-0844-3
15. Wang, J., Huang, Y.: Approach-avoidance pattern of attentional bias in individuals with high tendencies toward problematic Internet pornography use. Front. Psychiatry **13**, (2022). Art no. 988435. https://doi.org/10.3389/fpsyt.2022.988435

Non-Fungible Token Verification Scheme Based on Core Peripheral Sharding

Zijuan Chen[1,2], Jianyong Yu[1,2](✉), and Yulong Wang[1]

[1] Sanya Institute of Hunan University of Science and Technology, Sanya 572024, Hainan, China
chenhaha4086@163.com, yujyong@hnust.edu.cn, wyl@mail.hnust.edu.cn
[2] Hunan University of Science and Technology, Xiangtan 411201, Hunan, China

Abstract. As the fusion of the real world and the digital space, the metaverse is constructing a new interaction paradigm and economic system. Non-Fungible tokens (NFTs) have become a key technology to solving the digital asset ownership problem by their uniqueness and indivisibility. There are three core flaws in existing NFT verification schemes regarding privacy protection. First is the risk of identity exposure. Second, compatibility is insufficient. Traditional protocols rely on the homogenization property of currency, which is difficult to adapt to the indivisibility of NFT. Third, there is a trade-off between verification efficiency and privacy. These problems require new schemes to achieve anonymity, data confidentiality and resource consumption optimization while ensuring ownership verifiability.

To solve these challenges, we propose a privacy-preserving NFT scheme based on dynamic commitment and fragmentation architecture (NFTCP). Its innovation is reflected in three aspects: sharded Merkle tree architecture, a two-level cryptographic commitment mechanism, and a lightweight zero-knowledge proof circuit to verify commitments. Experiments show that the NFTCP is significantly better than the traditional NFT scheme regarding anonymity and confidentiality. Although the processing time of NFTCP is higher than that of ERC-721, its effective transaction ratio is increased by 106 percent than that of ERC-721 and 4.5 percent than that of other schemes, this cost is verified to be a reasonable compromise for privacy protection.

Keywords: Verifiable Security · Privacy Protection · Non-Fungible Token

1 Introduction

As a next-generation Internet form with deep integration of the physical world and digital space, the metaverse is reconstructing the communication mode and economic system of human society through the collaborative innovation of Extended Reality (XR), Digital Twin (DT) and blockchain. Blockchain-based Non-homogenous Tokens (NFTs) have become a key technology to solving the problem of digital object ownership of their uniqueness and indivisibility. The NFT standard represented by ERC-721 [1] realizes the casting and transfer of assets through smart contracts, but its public and transparent characteristics lead to severe challenges to user privacy. The blockchain address of the

J. S. Park et al. (Eds.): FutureTech 2025, LNEE 1483, pp. 130–139, 2025.
https://doi.org/10.1007/978-981-95-1999-6_17

owner in the transaction is directly exposed on the chain, and an attacker can reverse trace the user's identity through address association analysis. For example, studies [2] have shown that it takes only 3–5 transactions to associate an anonymous address with a real identity. This drawback is particularly prominent in sensitive scenarios such as virtual real estate transfers, which urgently need an innovative solution that considers both verifiability and privacy protection.

Current verification schemes face three core challenges in the field of privacy protection. First, the risk of identity exposure is significant. Mainstream NFT standards directly disclose the owner's address through the smart contract interface, and attackers can use heuristic address clustering and behaviors analysis to crack the user identity. Secondly, traditional privacy-preserving cryptocurrency schemes (such as Zcash [3] and ring signature [4]) rely on the homogeneity property and value conservation constraints of money, which are difficult to adapt to the non-homogeneity characteristics of NFT. For example, Zcash's validation logic needs to ensure that the input and output amounts of a transaction are equal, and the indivisible nature of NFTS makes this mechanism ineffective. Finally, the trade-off dilemma between efficiency and privacy limits practical applications. Although universal zero-knowledge proof (ZKP) can hide the details of transactions, its high computational complexity and on-chain storage overhead make it difficult to realize high-frequency trading scenarios. Existing privacy technologies are mainly designed around homogeneous cryptocurrencies, and there are significant limitations in their application in NFT scenarios. Monero [5] hides the address of the transaction sender through the coin mixing mechanism, but it relies on the divisibility of the transaction amount, which contradicts the indivisibility of NFT. Hawk protocol [6] requires the contract life cycle to be predefined, which cannot support the long-cycle circulation of NFTS. Specifically, the existing schemes lack adaptation to non-homogeneous characteristics, the verification efficiency grows linearly with the transaction volume, and the static Merkle tree structure is hard to cope with the expansion of transaction volume.

We proposed NFTCP to solve the above problems. Compared to other solutions, our work has the following uniqueness. The first is the sharded Merkle tree architecture, which uses dynamic sharding to manage the commitment tree. The core node maintains the global root hash and sequence number list, and the peripheral nodes perform lightweight verification. The second is the two-level encryption commitment mechanism, which binds the user's public key and random number through the first-level commitment, and the second-level commitment combines the token hash with the first-level result. Statistical concealment was used to prevent transaction association analysis, and the dynamic random number update strategy was introduced to resist historical data tracing. Thirdly, the ZKP circuit is customized to optimize the logic to verify the legitimacy of commitment, the consistency of ownership transfer and the uniqueness of sequence number. Experiments show that NFTCP achieves a better compromise between security and efficiency than other schemes and can effectively resist a series of security attacks. Figure 1 shows the scheme model diagram.

The subsequent chapters of this paper are organized as follows: The second chapter reviews relevant prerequisites. The third chapter describes the implementation details of the NFTCP scheme. In the fourth chapter, the effectiveness of the scheme is verified by informal security analysis and performance experiments. The fifth chapter summarizes the research conclusions and looks forward to future directions.

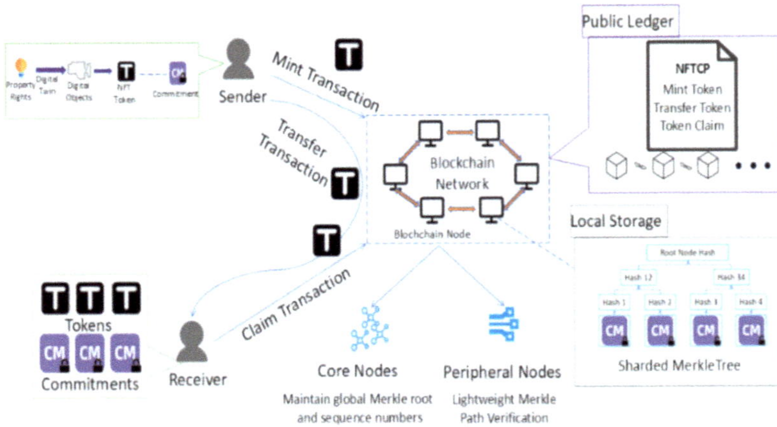

Fig. 1. NFT life cycle and scheme model diagram

2 Prerequisites

2.1 Core Periphery Identification

The identification of core-periphery structures has become an important research direction in network science. This structure division helps to reveal the basic organization principles of the network [7]. As the most densely connected key hub in the network, the core node undertakes the core functions of information dissemination and resource exchange. Peripheral nodes are mainly distributed at the edge of the network and only maintain limited local connections [8]. We innovatively apply the idea of core-periphery identification to the security assurance and work efficiency improvement of NFT verification schemes. The core node is responsible for protecting the security of transaction verification at the global level, and the peripheral nodes only need lightweight Merkle path verification, which effectively separates time-consuming cryptographic operations from constant level work processes.

2.2 Merkle Tree Verification

Merkle tree verification has been widely used as a data integrity verification mechanism based on the hash function [9]. The Merkle tree cleverly superimposes the hash values of multiple data blocks by constructing a complex tree structure. Each branch of the Merkle tree represents the integrity state of one or more specific blocks, and the root is the aggregated hash of all those blocks. All the verifier needs to do is provide a path from the destination data to the root hash.

2.3 Zero-Knowledge Proof

ZKP is an important cryptographic primitive that allows a prover to prove the truth of a statement to a verifier without revealing any additional information. ZKP has three key features: completeness, soundness, and zero-knowledge. ZKP has a wide range of applications in privacy protection, identity authentication and blockchain [10].

2.4 Cryptographic Commitment

Cryptographic commitment is a basic cryptographic primitive that allows a promising party to bind a secret value to a public commitment value while guaranteeing that the secret value will not be revealed until the appropriate time [11]. Cryptographic commitment plays a key role in numerous security protocols, such as zero-knowledge proofs and blockchains, especially in scenarios where fair exchange or verifiable delayed disclosure needs to be achieved. A secure commitment satisfies two properties:

- Hiding: Ensure that the commitment value does not leak any information about the secret value. Mathematically, for any two different secret values, v_1 and v_2, and the random number γ, we have:

$$P_r[C(v_1, r) = C(v_2, r)] \approx negl(\lambda) \tag{1}$$

  Here, C denotes the commitment function, λ is the security parameter, and $negl(\lambda)$ denotes the negligible probability.
- Binding: The promising party is guaranteed not to change the committed secret value after the fact.

3 Scheme Overview

3.1 Initialize Phase

The key generation algorithm is run by the authority to generate two pairs of keys K for NFT transaction and verification for both sides of the scheme. Where K contains the address keys and the encryption keys. The generated key will be distributed to participants through the smart contract. K is then used to generate and validate proofs in the corresponding transactions. In addition, the public keys PK_{addr} and PK_{enc} are published on the blockchain, and only the user knows the private key.

3.2 Mint Token Phase

Minting is the process of transforming digital content into a unique asset on the blockchain. In the blockchain technology system, the identification of digital objects is a key means of identity verification, which ensures the uniqueness and security of digital assets. By identifying the digital object, the user generates a unique identifier so that each digital asset has a unique identity. To achieve this goal, it is first necessary to preprocess the metadata. Users need to record the basic description information of digital objects and save it to a file, which further simplifies the management of NFT transaction data. After pretreatment, the random number injection is to enhance the key step in the metadata and uniqueness. The user must generate the random number and inject it as an additional attribute of the metadata. This step increases the randomness of the metadata and ensures that the tokens generated from the combination of metadata and random numbers have uniqueness. Finally, the hash value of the metadata was calculated by SHA-256 to obtain a unique token identifier *tID*. Each token consists of a *tID* and a description *meta*.

To protect the privacy and security of user identity information and tokens, NFTCP adopted a three-layer structure of a double commitment mechanism: random parameter generation, address binding commitment and token binding commitment. In the first layer, the user uses random numbers m, n and σ to generate unpredictable random parameters. These parameters are used to generate intermediate commitment and hide the user's public key to prevent the address information from being exposed. The intermediate commitment C_{mid} is calculated as follows.

$$C_{mid} = \sigma \cdot G + m \cdot H\left(PK^{A}_{addr}\right) + n \cdot G \tag{2}$$

where G is the generator, PK^{A}_{addr} is the user's public key, and H is the hash function. In the second layer, the token binding commitment binds the token identifier *tID* with the intermediate commitment C_{mid} using the random number s to form the final commitment C_f. The final commitment C_f is calculated as follows.

$$C_f = s \cdot G + tID \cdot H(C_{mid}) \tag{3}$$

This double commitment structure ensures the complete anonymity of token and user identity and effectively protects user privacy security. When the mint transaction$tx_{mint} =< meta, C_f, C_{mind}, s >$ is submitted to the blockchain network, the election process of the core node is triggered. Candidate nodes must pledge native tokens with a minimum threshold of $token_{min} = 1000$. If the candidate node does not meet this requirement or there is malicious behaviour trying to tamper with or attack the network, its pledged assets may be confiscated. The node weight is jointly calculated by the holding S_i and the historical verification success rate α_i (the initial value is 100 per cent). The introduced logarithmic function can suppress the resource monopoly and ensure the participation opportunity of small and medium-sized nodes. The node weight is calculated as follows.

$$weight_i = log(S_i) \times \alpha_i \tag{4}$$

After blocks are generated, the core nodes are elected as the new core verification group members, and the transparency and non-repudiation of the election process are ensured by Threshold Signature (TSS). In the threshold signature mechanism, the system uses a Distributed Key Generation Protocol (DKG) to generate corresponding private key shares $S = \{S_1, S_2, \ldots S_n\}$ and verifiable key set $V = \{V_1, V_2, , , V_n\}$ and has a public group public key X. The specific process of DKG protocol is as follows.

1. Initializing parameters: security parameter γ is chosen to determine the number of nodes n and the threshold t, where $t < n$. Each node p_i generates its own key pair (SK_i, PK_i), where SK_i is the private key and PK_i is the corresponding public key.

2. Polynomial construction: each node chooses a random degree $t-1$ polynomial $f_i(x) \epsilon Z_q[x]$ hat satisfies $f_i(0) = SK_i$. Then compute the values of this polynomial at the indices of the other nodes, that is, $f_{ij} = f_i(j) modq$ for each $j \neq i$.
3. Secret sharing distribution: p_i sends the calculated f_{ij} to node j_p. This process must ensure the confidentiality and integrity of the information.
4. Verification and reconstruction: p_j collects f_{ij} values from other nodes and verifies their legitimacy. If the verification passes, p_j can reconstruct the private key share sk_i of p_i using Lagrange interpolation.

$$SK_i = \sum_{j=1}^{n} f_{ij} \times \prod_{k=1}^{n} j/(j-k) modq(k \neq j) \tag{5}$$

   Here, j is the node index, and k is the interpolation point index.
5. Public key generation: all nodes cooperate to calculate the system's public key, which is calculated by the following equation.

$$PK = \sum_{i=1}^{n} PK_i modq \tag{6}$$

After the core node has verified the commitment C_f, the system updates the global state. The shard allocation is decided based on the hash value of the C_A to assign it to one of the local Merkle subtrees. The core node of each shard computes the local root hash independently to ensure the efficiency of parallel processing. Subsequently, each fragment root hash is aggregated into a global root through KZG polynomial commitment [12], and through the incremental synchronization strategy, peripheral nodes only need to synchronize the global root and the incremental path. This series of update operations enhances the blockchain network's stability and efficiency and improves the blockchain ecosystem's trust and scalability.

3.3 Transfer Token Phase

When user A tries to transfer the token he minted or received to user B, starts the transfer phase. To prove its ownership of a particular token, A first needs to generate a unique Serial Number (SN). This step requires user A to combine his private key SK^A_{addr} with the token commitment C_A, utilizing a hash function.

$$SN_A = H\left(C_A || SK^A_{addr}\right) \tag{7}$$

SN_A is permanently recorded on the blockchain to form a global sequence number list SN_{list}, ensuring it cannot be reused. Subsequently, A needs to generate random values and use them and PK^B_{addr} to calculate a new token commitment C_B. To enable B to verify A's token ownership, A uses symmetric encryption of the random number and tID to obtain the ciphertext Enc_{AB}. After completing the above process, A will construct proof π_A to verify the claim's validity without revealing private information. π_A consists of the following points:

- Commitment legitimacy: The commitments are generated by the user's public key along with the legitimate parameters, and they legally exist in the Merkle tree.

- Binding of sequence number: The SN_A is correctly generated by the user's private key and commitment.
- State consistency: The commitments are not marked as invalid (i.e., not included in the SN_{list}).

The proof is generated using Groth16 [13], which can generate concise proofs, ensuring both verification efficiency and privacy. Nodes complete the verification process by a predefined verification key. Eventually, A sends $MSG =< \pi_A, SN_A, C_B, Enc_{AB} >$ to the blockchain. The sharded Merkle tree updates are synchronized with the global state. The core nodes determine the shard it belongs to according to the hash value of C_B and obtains the subtree path $path_j$. Each shard core node computes the local root hash rt_j in parallel and aggregates the shard root hash into a global root rt_{global} by KZG polynomial commitment. Finally, peripheral nodes synchronize the global root and incremental path. After this, all nodes on the blockchain will verify the MSG sent by A. Specifically, they perform a check on whether proof, sequence number and commitment in MSG. If the MSG is verified by all nodes, the data in it is aggregated into transactions and added to the ledger and SN_A will also be added to SN_{list}. For each transfer transaction, B needs to decrypt the ciphertext in MSG using his encrypted private key SK^B_{Enc}. If the decryption is successful, Bob can use the decrypted information to compute a new commitment C_{B^*}. If C_{B^*} is the same as C_B in the transfer transaction, B will accept the token. The knowledge obtained by B after decryption can be further exploited in the claim of token ownership.

3.4 Token Claim Phase

In NFT transactions, users are required to make token ownership declarations to ensure that their ownership of digital assets is clearly documented and legally protected, while preventing infringement and protecting intellectual property rights. First, the user generates SN_C by combining their private key with the token commitment using a cryptographic hash function. SN_C is permanently recorded in SN_{list} to prevent duplicate claims. Next, the user constructs a proof π_C to verify the claim without revealing private information. Once the proof is constructed, blockchain nodes perform a verification process using a predefined verification key. Subsequently, the core node validates the existence of commitment in the sharded Merkle tree and retrieving the local subtree path $path_j$. $path_j$ is used to verify whether cm legally exists in the Merkle tree using rt_{global}. If the verification fails, the claim transaction is rejected. Upon successful verification, the core node updates the global state by adding SN_C to SN_{list} to prevent reuse. The sharded Merkle tree's local root hashes rt_j are aggregated into a global root, and peripheral nodes synchronize the global root and incremental path. Finally, the claim transaction is recorded on the blockchain ledger, and the user's ownership of the token is officially recognized. This phase ensures the privacy of token ownership claims. Figure 2 depicts the verification flow of NFTCP.

4 Scheme Analysis

This section provides a comprehensive evaluation of the three schemes, traditionally ERC-721, NFTPrivate [14], and NFTCP. The experimental environment is configured with a comprehensive setup including an Intel Core i5-12490F CPU, NVIDIA GeForce RTX 3060 Ti GPU, 16 GB RAM, and Ubuntu 20.04 LTS. Hyperledger Fabric v2.4 is deployed as the blockchain platform.

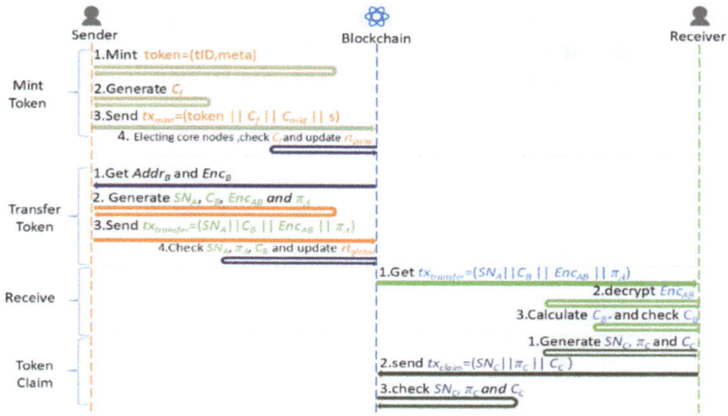

Fig. 2. The verification flow of NFTCP

4.1 Performance Evaluation

1. Valid Transactions

The number of valid transactions reflects the system's success rate in processing transactions. Figure 3 shows that ERC721 has a relatively low number of valid transactions, while the NFTPrivate and NFTCP schemes perform better in this regard. Specifically, the NFTCP scheme demonstrates a higher number of valid transactions across all transaction scales. For instance, at a scale of 1,000 transactions, NFTCP processes 750 valid transactions, compared to 725 for NFTPrivate. This trend continues with larger transaction scales.

2. Average Processing Time

Average processing time reflects the time required by the system to process a single transaction. At a scale of 1,000 transactions, the average processing time for NFTCP is 0.0005 s, while it is 0.00118 s for NFTPrivate. As the transaction scale increases, NFTCP's average processing time grows more slowly, indicating its efficiency in handling large-scale transactions.

4.2 Network Trustworthiness Analysis

Network trustworthiness is a key indicator of a system's reliability and stability. The distribution of invalid transaction reasons reflects the types and frequency of errors that may occur during transaction processing. NFTCP has significant advantages over NFT-Private in this area. At all transaction scales, NFTCP did not encounter any instances of invalid ZKPs. For example, at a scale of 1,000 transactions, NFTPrivate had 1 invalid ZKP, and this number increased to 6 at 50,000 transactions. In terms of invalid commitments, NFTCP also performs better. At a scale of 1,000 transactions, NFTCP had only 50 invalid commitments, while NFTPrivate had 177. Overall, NFTCP outperforms NFTPrivate in the distribution of invalid transaction reasons, including invalid ZKPs, commitments, and double-spending. This indicates that NFTCP is more reliable in NFT verification.

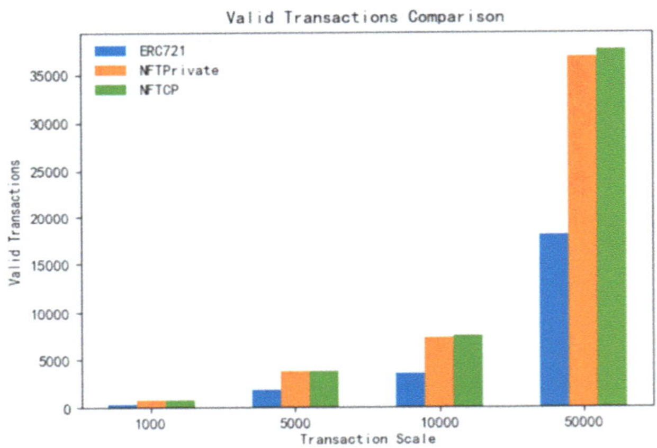

Fig. 3. Comparison of average processing time of different NFT verification schemes

4.3 Informal Safety Analysis

NFTCP employs a dual commitment mechanism with three cryptographic layers to ensure complete anonymity. Users generate random parameters combined with public keys and hash functions to create intermediate commitments, then bind token identifiers through randomized final commitments, preventing identity and token exposure throughout minting, transfer and claim processes.

5 Conclusion

This paper proposes a NFT transaction verification scheme based on core peripheral shard. The core verification node is selected through the dynamic proof of stake mechanism, the global state is innovatively decomposed into multiple local Merkle subtrees maintained in parallel, and the advanced cryptographic commitment technology is used to aggregate the fragmented data. Based on ensuring data integrity, the storage and verification overhead is significantly reduced. In the future work, we will further study how to achieve anonymous verification of NFT.

References

1. Averin, A., Samartsev, A.: Review of blockchain in computer games. In: AIP Conference Proceedings, vol. 2910(1). AIP Publishing (2023)
2. Banaeian Far, S., Hosseini Bamakan, S.M.: NFT-based identity management in metaverses: challenges and opportunities. SN Appl. Sci. **5**(10), 260 (2023)
3. Hopwood, D., Bowe, S., Hornby, T., et al.: Zcash protocol specification, vol. 4(220), p. 32. San Francisco, CA, USA, GitHub (2016)
4. Liu, Z., Gao, H., Lei, H., et al.: Overview of non-fungible token eco-regulation, traceability and data right confirmation security. In: International Conference on Information Science, Communication and Computing, pp. 95–109. Springer Nature Singapore, Singapore (2023)
5. Möser, M., Soska, K., Heilman, E., et al.: An empirical analysis of traceability in the monero blockchain. arXiv preprint arXiv:1704.04299 (2017)
6. Kosba, A., Miller, A., Shi, E., et al.: Hawk: the blockchain model of cryptography and privacy-preserving smart contracts. In: 2016 IEEE Symposium on Security and Privacy (SP), pp. 839–858. IEEE (2016)
7. Rombach, P., Porter, M.A., Fowler, J.H., et al.: Core-periphery structure in networks (revisited). SIAM Rev. **59**(3), 619–646 (2017)
8. Chen, Z., Yu, J., Wang, Y., et al.: Analysis of core-periphery structure based on clustering aggregation in the NFT transfer network. Entropy **27**(4), 342 (2025)
9. Jing, S., Zheng, X., Chen, Z.: Review and investigation of merkle tree's technical principles and related application fields. In: 2021 International Conference on Artificial Intelligence, Big Data and Algorithms (CAIBDA), pp. 86–90. IEEE (2021)
10. Hasan, J.: Overview and applications of zero knowledge proof (ZKP). Int. J. Comput. Sci. Network **8**(5), 2277–5420 (2019)
11. Wang, L., Shen, X., Li, J., et al.: Cryptographic primitives in blockchains. J. Netw. Comput. Appl. **127**, 43–58 (2019)
12. Zhang, J., Xie, T., Hoang, T., et al.: Polynomial commitment with a {One-to-Many} prover and applications. In: 31st USENIX Security Symposium (USENIX Security 22), pp. 2965–2982 (2022)
13. Gailly, N., Maller, M., Nitulescu, A.: Snarkpack: practical snark aggregation. In: International Conference on Financial Cryptography and Data Security, pp. 203–229. Springer International Publishing, Cham (2022)
14. Xiao, Y., Xu, L., Zhang, C., et al.: Blockchain-empowered privacy-preserving digital object trading in the metaverse. IEEE Multimedia **30**(2), 81–90 (2023)

Integrating TCM Constitution Theory with Deep Learning for Tongue Image Analysis

Zixuan Chen[1], Jin Wang[1,2], Yuzhen Liu[1], Xiaolan Zhou[1], and Xiaoliang Wang[1,2](✉)

[1] School of Computer Science and Engineering, Hunan University of Science and Technology, Xiangtan 411100, China
fengwxl@hunst.edu.cn

[2] Sanya Institute of Hunan University of Science and Technology, Sanya 572024, China

Abstract. This paper presents a comprehensive tongue diagnosis system that integrates Traditional Chinese Medicine (TCM) constitution theory with advanced deep learning techniques. Tongue diagnosis, a vital component of TCM's "inspection" method, has long relied on practitioners' subjective experience, leading to inconsistent diagnostic standards and susceptibility to external factors. To address these challenges, we propose a system combining YOLOv11 for detection, U2-Net for segmentation, and ResNet50 for classification of tongue images. Our system leverages the strengths of these models to overcome limitations in traditional tongue diagnosis, offering an efficient and accurate solution. We constructed a standardized dataset of 3,000 tongue images across nine constitutional types, collected under varied lighting conditions to ensure robustness. The experimental results demonstrate the system's effectiveness in automatically detecting, segmenting, and classifying tongue images, with significant improvements in diagnostic accuracy and efficiency. This study paves the way for the objectification and standardization of TCM inspection, providing a promising tool for mobile-based TCM diagnostic assistance.

Keywords: Tongue Diagnosis · Deep Learning · Traditional Chinese Medicine · YOLO · U2-Net · ResNet · Constitution Identification

1 Introduction

Tongue diagnosis, a critical component of the "inspection" method in the four diagnostic approaches of Traditional Chinese Medicine (TCM)—observation, auscultation and olfaction, inquiry, and palpation—has been practiced for thousands of years since the compilation of Huangdi Neijing (The Yellow Emperor's Inner Canon). By examining the color, shape, and coating of the tongue, practitioners can infer an individual's physiological state and pathological processes, playing a vital role in TCM's syndrome differentiation and treatment. However, traditional tongue diagnosis heavily relies on the subjective experience of practitioners, leading to inconsistent diagnostic standards and susceptibility to external factors such as lighting conditions, environmental factors, and visual fatigue, which may result in misjudgments and delayed treatment. For instance, as shown in Fig. 2, tongue features of three patients under different lighting conditions exhibit variations, potentially affecting diagnostic accuracy and treatment timing.

J. S. Park et al. (Eds.): FutureTech 2025, LNEE 1483, pp. 140–148, 2025.
https://doi.org/10.1007/978-981-95-1999-6_18

In recent years, the rapid advancement of artificial intelligence (AI) and image processing technologies has spurred research into the informatization and intelligent analysis of tongue images. Traditional machine learning methods, such as Support Vector Machines (SVM), Random Forests (RF), and K-Means clustering, have been applied to extract and classify color, texture, and shape features of tongue images. While these methods have achieved some success, they suffer from limitations in handling complex features, ensuring model stability, and resisting environmental interference. Additionally, most approaches still require manual assistance, which fails to meet the demands for efficiency and objectivity in clinical applications.

The introduction of deep learning techniques has brought significant breakthroughs to tongue image recognition. Models such as the YOLO series, Faster R-CNN, U^2-Net, and ResNet have demonstrated exceptional performance in tasks like object detection, semantic segmentation, and feature classification, and have been widely adopted for tongue image analysis. However, existing studies primarily focus on classifying specific tongue features, such as cracked or tooth-marked tongues, and lack a comprehensive system grounded in TCM's "constitution identification" theory.

This study integrates TCM constitution theory to construct a standardized training dataset of tongue images representing nine constitutional types: Balanced Constitution, Qi Deficiency Constitution, Yang Deficiency Constitution, Yin Deficiency Constitution, Phlegm-Dampness Constitution, Damp-Heat Constitution, Blood Stasis Constitution, Qi Depression Constitution, and Hereditary Constitution. We propose a comprehensive tongue diagnosis system that combines YOLO for detection, U^2-Net for segmentation, and ResNet for classification, enabling automatic detection, precise segmentation, and constitution classification of tongue images. The overall workflow of the proposed system is illustrated (see Fig. 1).

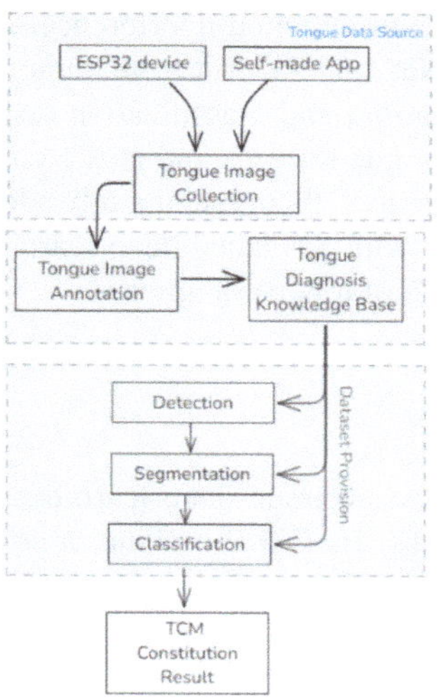

Fig. 1. Tongue Image Analysis Workflow

This method not only enhances the efficiency and accuracy of tongue diagnosis but also provides a technical foundation for the objectification and standardization of TCM inspection. By leveraging advanced deep learning models, our system aims to overcome the limitations of traditional tongue diagnosis and offers a promising solution for mobile-based TCM diagnostic assistance systems.

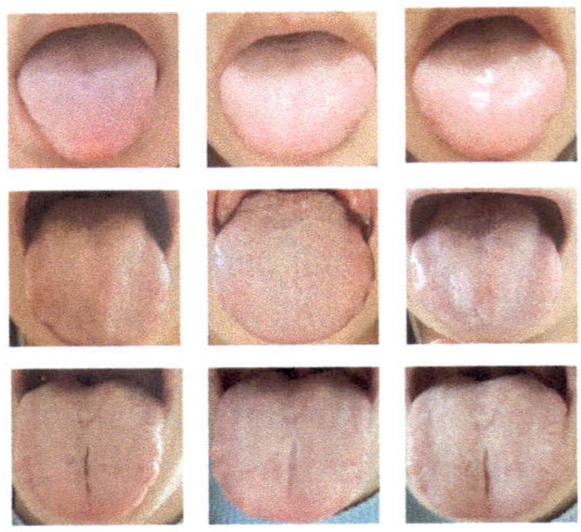

Fig. 2. Comparative Study of Tongue Images Under Different Lighting Conditions

2 Related Work

Traditional tongue diagnosis in TCM relies heavily on practitioners' subjective experience, leading to inconsistent diagnostic standards and susceptibility to external factors. Recent advancements in AI and image processing have spurred research into tongue image analysis. Traditional machine learning methods like SVM, RF, and K-Means have been applied for feature extraction and classification but face limitations in handling complex features and environmental interference. Deep learning models such as YOLO, Faster R-CNN, U^2-Net, and ResNet have demonstrated superior performance in object detection, segmentation, and classification tasks. However, existing studies primarily focus on specific tongue features and lack a comprehensive system based on TCM's "constitution identification" theory. Our study addresses these gaps by integrating TCM constitution theory with advanced deep learning techniques to enhance the objectivity and standardization of tongue diagnosis.

3 Methods

YOLOv11 is a high-efficiency, accurate deep learning-based object detection model widely used for real-time tasks. Its C3K2 module, a critical and flexible component, integrates the advantages of the C2F module and allows dynamic adjustments via tunable parameters.

3.1 YOLO: A Deep Learning-Based Object Detection Model

YOLO-C3K2

When the c3k parameter is FALSE, it uses a traditional Bottleneck design for speed-prioritized tasks, though with limitations in complexity. When TRUE, it replaces the Bottleneck with the C3 module, enhancing feature fusion and contextual modeling for precise object localization in complex scenarios. This adaptable design enables YOLOv11 to balance efficiency and accuracy, making it a versatile framework for diverse detection tasks (Fig. 3).

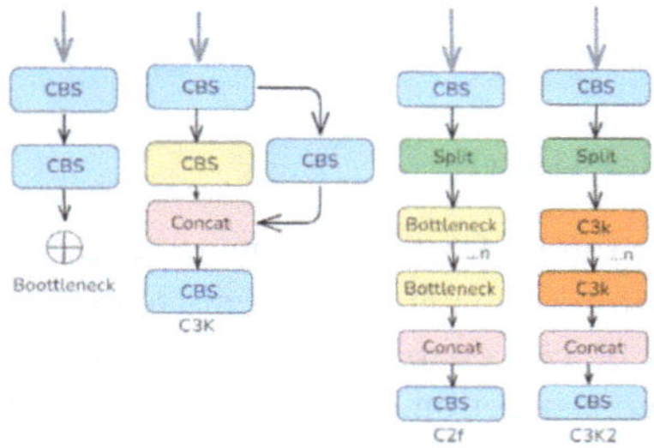

Fig. 3. The network architecture of C3K2.

The C2PSA Module

The C2PSA module enhances the C2F module by integrating a PSA (Pointwise Spatial Attention) block, which improves feature extraction and introduces a stronger attention mechanism. While the C2F module provides robust feature extraction, it struggles with complex backgrounds and small objects. The C2PSA module addresses this by focusing attention on key regions, enabling the model to better identify discriminative image parts and improve detection accuracy.

The PSA block employs a spatial attention mechanism at the pixel level, allowing the network to weigh important features more effectively. Unlike traditional convolutions, it adjusts spatially to emphasize critical regions like edges and textures while suppressing background noise. This enables superior performance in complex images and scenarios with small objects or dense backgrounds. By incorporating the PSA block, the C2PSA module retains the efficiency of the C2F module while enhancing local feature capture, achieving higher accuracy without significant computational overhead. This makes it a crucial component in YOLOv11, improving adaptability and robustness in real-world applications (Fig. 4).

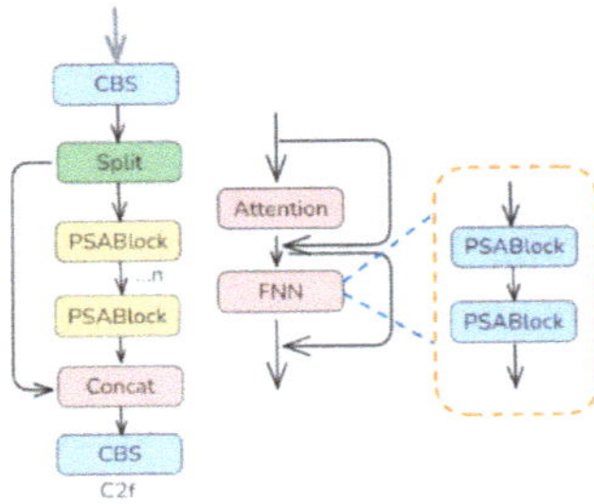

Fig. 4. Network architecture of C2PSA

3.2 Training of the U^{2}Net Model

This paper uses the U^{2}Net model for tongue image segmentation, leveraging its innovative encoder-decoder architecture and multi-scale feature extraction mechanism. The model's nested U-shaped blocks enable comprehensive feature extraction and high-resolution feature map reconstruction, providing strong edge-aware capabilities for precise segmentation of tongue features such as contours, coating, cracks, and tooth marks.

U^{2}Net incorporates deep supervision to address gradient vanishing issues, accelerating convergence and stabilizing training. Training utilized labeled tongue images and binary masks, combined with data augmentation strategies like rotation, flipping, and brightness adjustment, to enhance adaptability to diverse conditions.

After optimization, U^{2}Net demonstrated accurate tongue region segmentation and detailed contour delineation, supporting subsequent analysis and TCM diagnostic model development (Fig. 5).

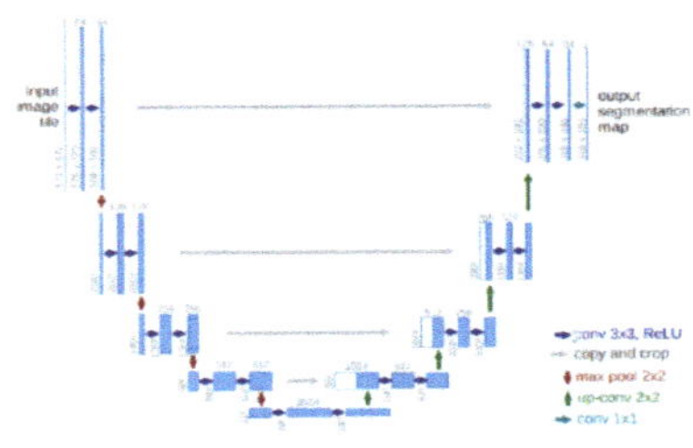

Fig. 5. U^{2}Net Model

3.3 Training of ResNet50 and Lightweight CNN Models

This study uses ResNet50 and a custom lightweight CNN for tongue feature extraction and constitution classification. Training leveraged diverse datasets from CAS and Shanghai University of TCM to enhance generalization. ResNet50's residual learning mechanism effectively extracts complex features, addressing gradient issues. The lightweight CNN reduces parameters for efficient mobile deployment without sacrificing accuracy.

Training included data augmentation and hyperparameter tuning to boost robustness and convergence.

4 Experiments

Table 1. Environments.

Category	Details
Operating System	Centos 7.9
GPU Model	NVIDIA GeForce RTX 3080
RAM	16 GB
Language	Python
IDE / Compiler	Pycharm
CUDA Version	12.1
PyTorch	2.3.0

4.1 Dataset

The dataset used in this study consists of 3,000 tongue images collected under three lighting conditions: backlight, natural light, and flash, with 1,000 images per condition. It was split into training (2,250 images) and validation sets (750 images) in a 3:1 ratio. Images captured under backlight conditions may show distortion due to illumination angles. Natural light images can have obstructions, curved shapes of small objects, and low brightness as interfering factors. Flash-condition images were taken in low-light environments with a flash to ensure visibility (Table 1).

4.2 Evaluation Metrics

The evaluation metrics employed in this study include Precision, Recall, and mean Average Precision (mAP).

Precision: is a key evaluation metric in classification tasks that measures the proportion of true positive predictions among all instances predicted as positive, reflecting the model's ability to accurately identify relevant cases.

$$\text{Precision} = \frac{\text{TP}}{\text{TP} + \text{FP}} \tag{1}$$

Recall: In classification tasks, Recall measures the proportion of actual positive instances that are correctly identified by the model, reflecting its ability to capture relevant cases comprehensively

$$\text{Recall} = \frac{\text{TP}}{\text{TP} + \text{FN}} \tag{2}$$

mAP: Mean Average Precision (mAP) is a comprehensive metric that evaluates the overall performance of a model across multiple classes by averaging the precision values at different recall thresholds, providing a balanced assessment of both precision and recall.

$$AP = \sum_{i=1}^{n-1} (r_{i+1} - r_i) * p(r_{i+1}) \tag{3}$$

Confusion Matrix: a tabular representation used to evaluate the performance of classification models by illustrating the relationship between predicted and actual class labels (Figs. 6 and 7).

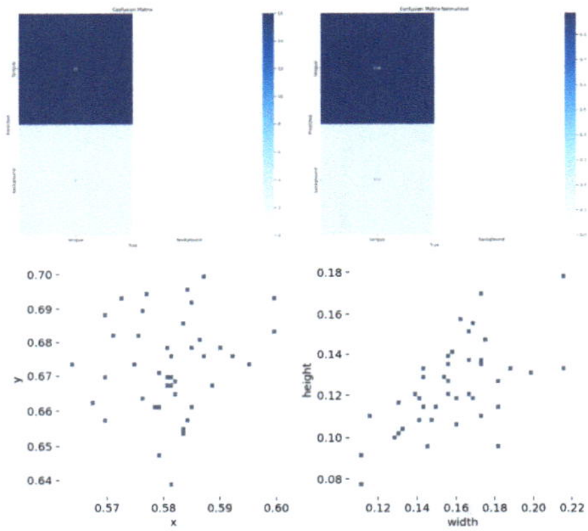

Fig. 6. Experimental Results Figure

4.3 Results

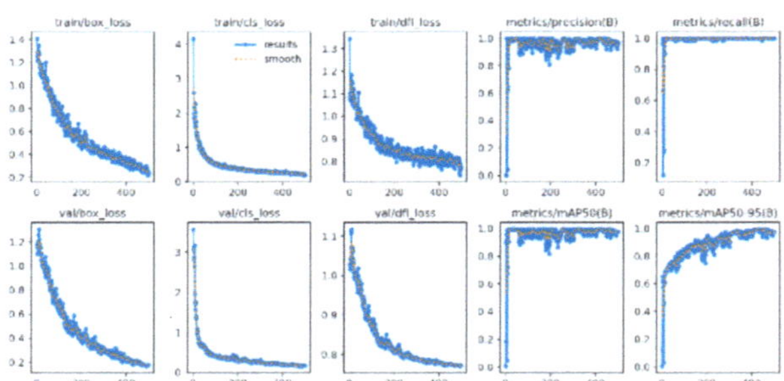

Fig. 7. Results

4.4 Visualization Results of Tongue Diagnosis Experiments

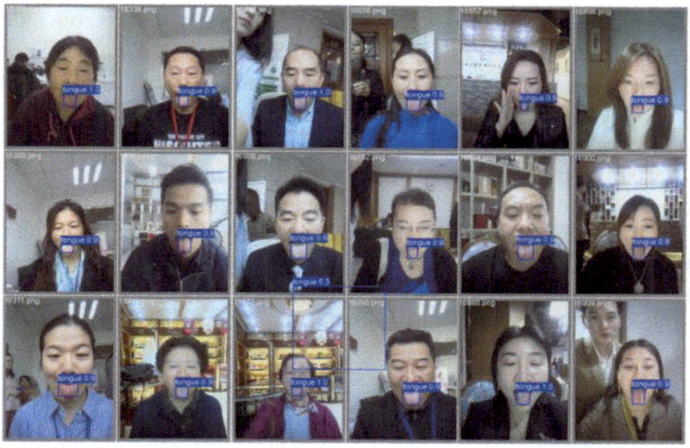

5 Future Work

In future research, we aim to further enhance the performance and applicability of our tongue diagnosis system. First, we plan to expand our dataset by collecting more tongue images under diverse environmental conditions and from a broader population to improve the model's generalization ability. Second, we intend to explore advanced deep learning architectures and optimization techniques to further improve the accuracy and efficiency of tongue feature extraction and classification. Third, we aim to integrate multimodal data, such as patient history and clinical symptoms, with tongue image analysis to provide a more comprehensive diagnostic support system. Additionally, we plan to develop a user-friendly mobile application to facilitate the practical deployment of our system in clinical settings, making TCM diagnosis more accessible and efficient. Finally, we will conduct more extensive validation studies to assess the clinical effectiveness and reliability of our system in real-world applications.

References

1. Cao, M.L., Zhang, X.F., Shen, L.S.: Application survey of information combination in the toughness and tenderness of tongue manifestation recognition. Beijing Biomed. Eng. **25**(6), 644–648 (2006)
2. Shi, Z., Zhou, C.L.: Fissure extraction and analysis of image of tongue. Comput. Technol. Develop. **17**(5), 245–248, 253 (2007)
3. Zhu, M.L.M., Lu, P., Xia, C.M., et al.: Research on douglas-peucker method in feature extraction from 55 cases of tooth-marked tongue images. Chin. Arch. Tradit. Chin. Med. **32**(9), 2138–2140 (2014)
4. Wang, X.M., Wang, R.Y., Guo, D., et al.: A research about tongue-prickled recognition method based on auxiliary light source. Chin. J. Sens. Actuat. **29**(10), 1553–1559 (2016)
5. Guo, D.: The research of tongue objectiveness based on binocular stereo vision. Tianjin University, Tianjin (2018)
6. Girshick, R., Donahue, J., Darrell, T., et al.: Rich feature hierarchies for accurate object detection and semantic segmentation. In: Proceedings of the IEEE Conference on Computer Vision and Pattern Recognition, pp. 580–587. IEEE, Columbus (2014)

7. He, K., Zhang, X., Ren, S., et al.: Spatial pyramid pooling in deep convolutional networks for visual recognition. IEEE Trans. Pattern Anal. Mach. Intell. **37**(9), 1904–1916 (2015)
8. Tang, W., Gao, Y., Liu, L., et al.: An automatic recognition of tooth-marked tongue based on tongue region detection and tongue landmark detection via deep learning. IEEE Access **8**, 153470–153478 (2020)
9. Akram, N., et al.: Exploiting the multiscale information fusion capabilities for aiding the leukemia diagnosis through white blood cells segmentation, vol. 10, pp. 48747–48760
10. Chen, H., He, Y: Machine learning approaches in traditional chinese medicine: a systematic review. **50**(1), 91–131
11. Deng, F., Lv, J.H., Wang, H.L., Gao, J.M., Zhou, Z.L.: Expanding public health in china: an empirical analysis of healthcare inputs and outputs, vol. 142, pp. 73–84

Intelligent Pulse Diagnosis in Traditional Chinese Medicine: Signal Processing and Deep Learning for Health Status Classification

Zixuan Chen[1], Jin Wang[1,2], Yuzhen Liu[1], Xiaolan Zhou[1], and Xiaoliang Wang[1,2(✉)]

[1] School of Computer Science and Engineering, Hunan University of Science and Technology, Xiangtan 411100, China
fengwxl@hunst.edu.cn

[2] Sanya Institute of Hunan University of Science and Technology, Sanya 572024, China

Abstract. The intelligent pulse diagnostic instrument, developed by Jianwei TCM Studio, integrates signal processing and deep learning for standardized pulse wave acquisition and recognition. This device combines hardware, including a sensor, pressure device, signal amplification module, and STM32 microcontroller, with software featuring Alibaba Cloud servers, HVG networking, and an attention-based ACRNN for pulse wave analysis and physique recognition. It offers two primary functions: standardized pulse wave data acquisition and cloud storage, enabling health monitoring through a user-friendly, portable design; and pulse wave preprocessing and physique recognition using HVG and ACRNN, achieving high diagnostic accuracy for sub-healthy populations. A prototype has been successfully developed with all functional modules operational.

Keywords: Intelligent pulse diagnostic device · Pulse wave · Signal processing · Deep learning · Convolutional recurrent neural network

1 Introduction

Pulse diagnosis, a fundamental component of Traditional Chinese Medicine (TCM), has been practiced for thousands of years as part of the "palpation" method in TCM's four diagnostic approaches. By analyzing the rhythm, strength, and depth of the pulse, practitioners can infer an individual's physiological and pathological states. However, traditional pulse diagnosis is highly subjective, relying on the practitioner's tactile experience and diagnostic expertise. This subjectivity leads to inconsistent diagnostic standards and potential misjudgments, especially under varying environmental conditions.

In recent years, the integration of modern technologies such as signal processing and deep learning has driven research into the objective and intelligent analysis of pulse signals. Traditional methods, including Fourier Transform and Wavelet Transform, have been used to extract pulse wave features. While these methods have achieved some success, they often fail to handle complex temporal and spatial patterns in pulse signals.

J. S. Park et al. (Eds.): FutureTech 2025, LNEE 1483, pp. 149–156, 2025.
https://doi.org/10.1007/978-981-95-1999-6_19

Deep learning models, particularly Convolutional Neural Networks (CNNs) and Recurrent Neural Networks (RNNs), have shown promise in capturing intricate patterns in pulse data.

This study introduces an intelligent pulse diagnostic system that combines signal processing with deep learning to achieve automated pulse signal acquisition, preprocessing, and constitution classification. The system leverages a novel attention-based Convolutional Recurrent Neural Network (ACRNN) to achieve high-precision pulse feature extraction and health status classification. The proposed framework not only enhances diagnostic accuracy but also provides a foundation for the standardization and digitization of TCM pulse diagnosis.

2 Related Work

Previous research in pulse diagnosis has focused on signal acquisition, feature extraction, and pattern recognition. Early studies utilized traditional signal processing techniques such as Fourier Transform and Wavelet Transform to analyze pulse waveforms. More recently, machine learning methods, including Support Vector Machines (SVM) and Random Forests (RF), have been applied to classify pulse features. However, these methods often require manual feature engineering and struggle with complex temporal dependencies in pulse signals.

Deep learning has emerged as a powerful tool for pulse signal analysis. Models such as CNNs and RNNs have been used to extract spatiotemporal features from pulse waveforms. For example, Long Short-Term Memory (LSTM) networks have been employed to model temporal dependencies in the data, while CNNs have been used to extract spatial features. Despite these advances, existing studies lack a comprehensive framework that integrates TCM's theoretical foundations with modern computational techniques.

Our work addresses this gap by proposing a unified framework that combines HVG network analysis for feature extraction and ACRNN for classification, providing a novel approach to pulse diagnosis.

3 Methods

3.1 Pulse Signal Acquisition and Preprocessing

The pulse diagnostic system employs a high-precision piezoelectric sensor (SDT1-028K) to capture pulse waveforms. The sensor is placed on the radial artery of the wrist, converting mechanical pulse waves into electrical signals. These signals are amplified using an AD620 instrumentation amplifier and filtered to remove noise. The preprocessing pipeline includes:

Filtering: Hig. :h-pass and low-pass filters to remove environmental noise and artifacts.

Normalization: Scaling the signal to a standardized range for consistency.

Segmentation: Dividing the continuous signal into discrete pulse cycles using peak detection.

3.2 HVG Network Analysis

The Horizontal Visibility Graph (HVG) is a powerful tool for transforming time series data into complex networks, providing a new perspective for analyzing dynamic systems. In the context of pulse wave analysis, HVG offers several advantages over traditional methods by capturing nonlinear and complex patterns inherent in pulse signals.

Principle of HVG

The HVG method, proposed by Lacasa et al. in 2008, is a variant of the Visibility Graph (VG) method. It simplifies the network construction process by restricting the visibility criterion to the horizontal direction. In this method, each data point in a time series is treated as a node in a network. A node at time *ti* is connected to another node at time *tj* if the following horizontal visibility condition is met:

$$y_i > y_k \text{ for all } t_i < t_k < t_j \tag{1}$$

This condition ensures that a horizontal line can be drawn between the two points without intersecting any intermediate data points. The resulting network structure preserves key characteristics of the original time series, such as trends, periodicity, and randomness.

Application to Pulse Wave Analysis

When applied to pulse wave analysis, the HVG method involves the following steps:

Data Preprocessing: The Pulse Wave Signal is First Preprocessed to Remove Noise and Artifacts. This typically includes filtering and normalization to ensure the signal is suitable for network construction.

Network Construction: The Preprocessed Pulse Wave Data is Converted into an HVG Network. Each peak in the pulse wave is identified and used as a node in the network. The horizontal visibility rule is applied to establish connections between these nodes.

Network Characterization: The Constructed HVG Network is Analyzed to Extract Various Topological Features. Key metrics include:

Degree Distribution: Reflects the connectivity of nodes and indicates the presence of hubs or highly connected nodes.

Clustering Coefficient: Measures the tendency of nodes to form clusters, providing insights into the network's local structure.

Network Diameter: : The longest shortest path between any two nodes, indicating the network's global connectivity.

Physiological Interpretation: These network metrics are then correlated with physiological states. For example, a pulse wave network with a scale-free degree distribution may indicate a healthy state, while deviations from this pattern could suggest pathological conditions.

Advantages of HVG in Pulse Wave Analysis

Nonlinear Insight: HVG captures nonlinear features that traditional time and frequency domain analyses might miss.

Dimensionality Reduction: Converts high-dimensional time series data into a lower-dimensional network representation, simplifying analysis.

Robustness: The method is robust to noise and artifacts, making it suitable for real-world physiological signals.

By integrating HVG with advanced machine learning models like the Attention-based Convolutional Recurrent Neural Network (ACRNN), we can achieve high-precision pulse wave analysis and health status classification. This combination not only enhances diagnostic accuracy but also provides a foundation for the standardization and digitization of Traditional Chinese Medicine pulse diagnosis.

3.3 ACRNN for Constitution Classification

The ACRNN model integrates channel attention and self-attention mechanisms to enhance feature extraction and classification. The model architecture includes:

Channel Attention: Dynamically assigns weights to different channels based on their importance.

Convolutional Layers: Extract spatial features from pulse waveforms.

LSTM Layers: Model temporal dependencies in the data.

Self-attention: Re-encodes sample importance based on similarity.

The ACRNN model is trained using a dataset of pulse waveforms collected from healthy and sub-healthy individuals. The training process optimizes the model to classify pulse signals into health categories with high accuracy.

4 Experiments

On the basis of analyzing the current research status of pulse waves, the VG network method has been introduced into pulse wave analysis. The pulse wave network constructed using the HVG method exhibits scale-free properties but lacks the characteristics of a small-world network. This complex network analysis approach expands the research perspective on pulse waves, enabling researchers to obtain pulse wave feature parameters from more dimensions.

4.1 Feasibility

The HVG network method is a simplified time-series network approach derived from the VG network construction method while preserving the network's properties. This method restricts the visibility criteria of VG, defining the relationship between data or nodes as the visibility between bars in a histogram. The rule for network construction is as follows: For any two data points A(ta, ya) and B(tb, yb), they have a mutual visibility relationship if any data point C(tc, yc) between them satisfies the following condition:

$$y_c < y_b + (y_a - y_b)\frac{t_b - t_c}{t_b - t_a} \tag{2}$$

The data point C(tc, yc) is located below the line connecting data points A(ta, ya) and B(tb, yb). When this condition is met, the corresponding nodes in the network are connected by an edge (Fig. 1).

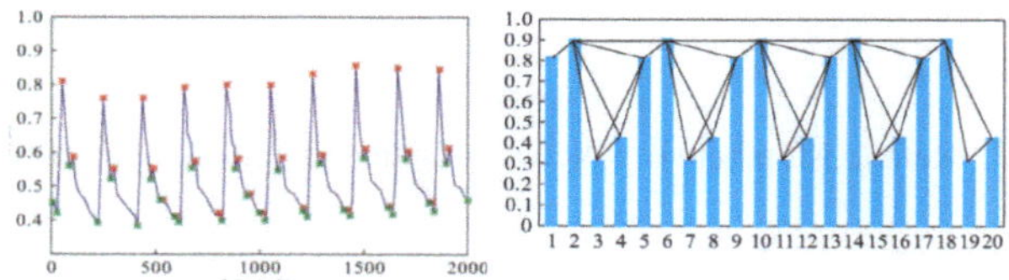

Fig. 1. Relationship between Data Points and Sampling Number

Based on the analysis of pulse waves, the periodic sequence of pulse waves is used as network input data. A new periodic automatic identification algorithm is employed to identify the peak positions of pulse waves, as shown in Figs. 4 and 3. In the figure, green markers indicate minima, red markers indicate identified maxima, and black markers indicate the filtered peak positions of pulse waves.

Practical Application Degree Distribution

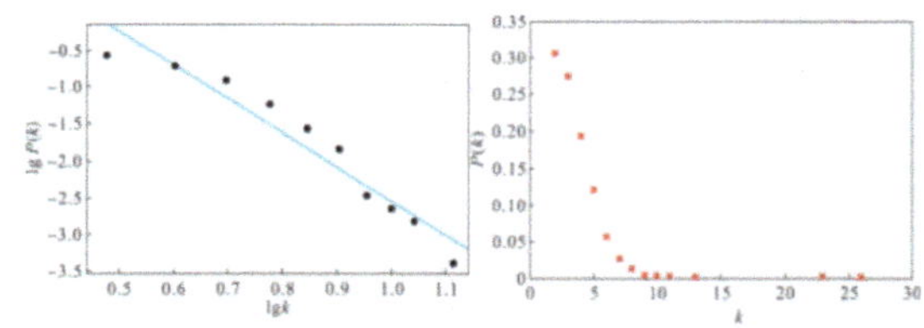

Fig. 2. Distribution of Real-World Applicability

Based on the analysis of pulse waves, the periodic sequence of pulse waves is utilized as network input data. A novel periodic automatic identification algorithm is employed to identify the peak positions of pulse waves, as shown in Figs. 4 and 3. In the figure, green markers indicate minima, red markers indicate identified maxima, and black markers indicate the filtered peak positions of pulse waves (Fig. 2).

Practical Application Degree Distribution

The fitting effect of the relationship between network diameter and network size is illustrated in Figs. 4, 5 and 6. Experimental data from a volunteer's pulse wave samples, collected over incrementally increasing time intervals, were used to analyze how the network diameter changes with network size. The results indicate a linear relationship between network diameter and network size, as evidenced by a high goodness of fit (0.8741). This suggests that the pulse wave network does not exhibit small-world characteristics, as the network diameter increases linearly with network size. This finding is

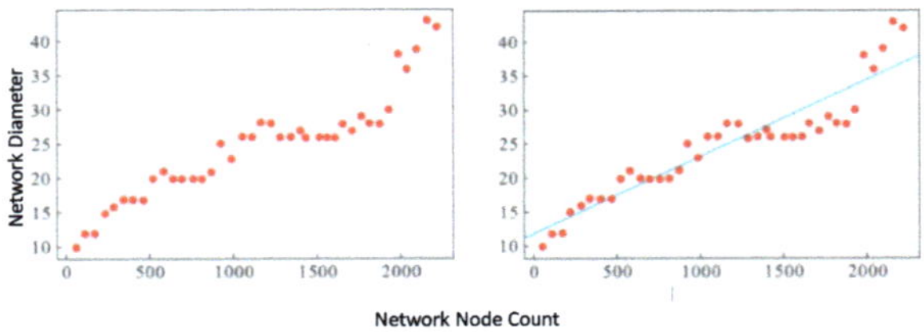

Fig. 3. Relationship between Network Diameter and Number of Nodes

significant for understanding the structural properties of pulse wave networks and their implications for physiological analysis

Lesion Analysis Algorithm

The Fourier Transform is a mathematical tool that converts time-domain signals into frequency-domain signals. In the context of medical imaging analysis, it is widely used to extract detailed structural information and global semantic features from signals. The discrete Fourier Transform (DFT) is defined as (Table 1):

$$X(k) = \sum_{n=0}^{N-1} x(n) \cdot e^{-2\pi ikn/N} \text{ for } k = 0,1, \ldots, N-1 \tag{3}$$

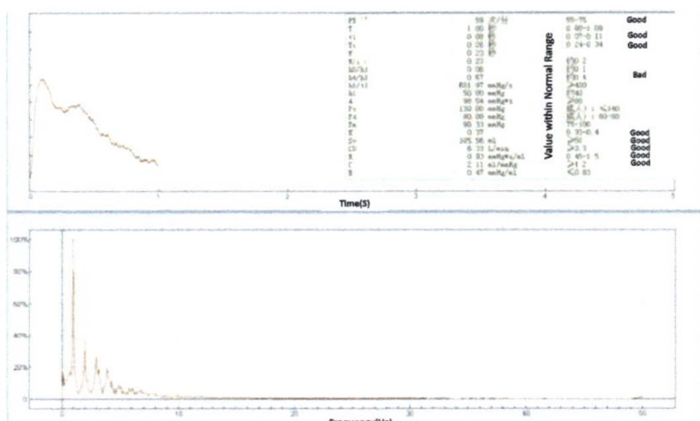

Fig. 4. Haar Wavelet

The Loss Function is as Follows

Table 1. Energy Amplitude Ratio of Related Organs.

Related Organ	Energy Amplitude Ratio (%)
Heart	100.000000
Liver	80.388410
Kidney	53.670830
Spleen	41.030000
Lung	22.520010
Stomach	17.643980

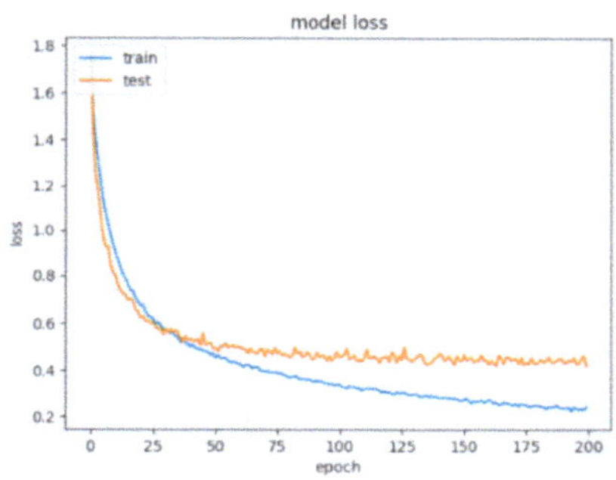

Fig. 5. Loss over Epochs

5 Future Work

Future research will focus on expanding the dataset to include diverse populations and environmental conditions. We plan to explore advanced deep learning architectures, such as Transformer-based models, to further enhance feature extraction and classification accuracy. Additionally, we aim to develop a mobile application for practical deployment, enabling real-time pulse diagnosis and telemedicine applications.

References

1. Tao, W., Li, C., Song, R., et al.: EEG-based emotion recognition via channel-wise attention and self attention. IEEE Trans. Affect. Comput. (2020)
2. Zhang, D., Yao, L., Zhang, X., Wang, S., Chen, W., Boots, R.: Eegbased intention recognition from spatio-temporal representations via cascade and parallel convolutional recurrent neural networks, arXiv preprint arXiv:1708.06578 (2017)
3. Hochreiter, S., Schmidhuber, J.: Long short-term memory. Neural Comput. **9**(8), 1735–1780 (1997)
4. Chen, L., et al.: Sca-cnn: spatial and channel-wise attention in convolutional networks for image captioning. In: Proceedings of the IEEE Conference on Computer Vision and Pattern Recognition, pp. 5659–5667 (2017)

5. Shen, T., Zhou, T., Long, G., Jiang, J., Pan, S., Zhang, C.: Disan: directional self-attention network for RNN/CNN-free language understanding. In: Thirty-Second AAAI Conference on Artificial Intelligence (2018)
6. ¨Ozerdem, M.S., Polat, H.: Emotion recognition based on EEG features in movie clips with channel selection. Brain Inform. **4**(4), 241 (2017)
7. Tong, L., Zhao, J., Fu, W.: Emotion recognition and channel selection based on EEG signal. In: 2018 11th International Conference on Intelligent Computation Technology and Automation (ICICTA), pp. 101–105. IEEE (2018)
8. Vaswani, A., et al.: Attention is all you need. In: Advances in Neural Information Processing Systems, pp. 5998–6008 (2017)

KITE-MRE: Knowledge-Infused and Transformer-Enhanced Multimodal Relation Extraction

Junjie Li(✉), Wenti Huang, Yu He, and Xinjie Mo

Hunan University of Science and Technology, Xiangtan 411201, Hunan, China
lijunj@mail.hnust.edu.cn

Abstract. Multimodal Relation Extraction (MRE) is a task that identifies semantic relations between two entities from image-text pairs. Existing methods often suffer from insufficient extraction accuracy and low cross-modal fusion efficiency due to inadequate auxiliary information. In this paper, we propose a multimodal relation extraction framework KITE-MRE,which integrating large model generation and knowledge graph (KG) to provide abundant additional auxiliary information. Specifically, prompts guide large language models to construct image semantic entity triplets and generate image graphs, while knowledge graph entity linking technology is applied in text graphs to build a unified graph structure. Additionally, we introduce the Graph Information Bottleneck (GIB) technique combined with Graph Convolutional Networks (GCN) to optimize, and perform feature fusion by integrating semantic and structural multi-perspective information to predict relations. Experimental results demonstrate that this approach provides a new path for multimodal data processing, effectively addressing issues of knowledge graph incompleteness and ambiguous relation generation by large models in cross-modal scenarios.

Keywords: Multimodal Relation Extraction · Large Language-Models · Knowledge Graph

1 Instruction

Multimodal Relation Extraction (MRE) aims to identify entity relationships by leveraging visual-textual complementarity. Traditional methods, relying on single-modal inputs or shallow cross-modal fusion, struggle to integrate heterogeneous data effectively. Recent advancements (e.g., MEGA's (Zheng et al. 2021) syntactic-visual graph alignment, ViLBERT's (Lu et al. 2019) cross-modal attention, HVPNet's (Chen et al. 2022) hierarchical fusion) partially address these challenges but remain limited by semantic fragmentation and insufficient external knowledge integration.

The emergence of Large Language Models (LLMs) and Knowledge Graphs (KGs) offers new opportunities. Recent work (Wang et al. 2022) demonstrates that LLM-KG synergy enhances relation extraction accuracy and recall, while (Chen et al. 2024) mitigates long-tail reliability issues via Graph Convolutional Networks (GCNs) and Graph Information Bottleneck (GIB).

J. S. Park et al. (Eds.): FutureTech 2025, LNEE 1483, pp. 157–164, 2025.
https://doi.org/10.1007/978-981-95-1999-6_20

This paper proposes the KITE-MRE framework, integrating large language models (LLMs) and knowledge graphs (KGs) for enhanced MRE. Key innovations include:

- *Unified Graph Construction:* LLM-assisted conversion of text and images into structured graphs, fused with KGs via entity linking
- *KG Subgraph Integration:* Addressing logical biases and improving reliability in long-tail relations through prior knowledge infusion.
- *Dynamic Feature Filtering*: Graph Information Bottleneck (GIB) removes redundant nodes/edges, while Graph Convolutional Networks (GCNs) enable efficient multimodal fusion.

2 Related Work

2.1 Multimodal Relation Extraction

Multimodal Relation Extraction (MRE) enhances entity relation classification through visual-text collaborative modeling. The MNRE dataset (Zheng et al. 2021), aligns visual object relations with textual entity relations for cross-modal evaluation. The MEGA framework further models images as scene graphs and text as syntactic trees, demonstrating graph-structured advantages in capturing cross-modal dependencies. Recent work extends MRE to Joint Multimodal Entity-Relation Extraction (JMERE), integrating edge alignment mechanisms for improved joint modeling (Yuan et al. 2023). Transformer-based architectures dominate due to cross-modal interaction capabilities, such as subword-level alignment for noise suppression (Li et al. 2023). To address modality heterogeneity, HVPNet dynamically integrates visual features via pyramid maps and gating attention (Xiang et al. 2022).

2.2 Knowledge Graphs and Large Language Models

Integrating Knowledge Graphs (KGs) with Large Language Models (LLMs) enhances structured reasoning. Early approaches use KGs as knowledge repositories: "Think-on-Graph" retrieves triples for QA tasks (Sun et al. 2023), while "Reasoning on Graph" generates multi-hop paths for LLM guidance (Luo et al. 2023). To improve credibility, methods like "Chain-of-Knowledge" combine KGs with heterogeneous sources to reduce hallucinations (Li et al. 2024).

2.3 Graph Structure Optimization: Information Bottleneck Technique

The Information Bottleneck (IB) theory (Alemi et al. 2017) compresses data by retaining critical information. For graphs, Graph IB (GIB) (Wu et al. 2020) refines graph G into a compact representation G^- by minimizing mutual information between G and G^-, while ensuring G^- sufficiently predicts labels Y.

3 Method

3.1 Framework Overview

The proposed framework (Fig. 1) consists of four stages: Feature Encoding, where a Transformer-based language model extracts semantic features from text and images; Graph Construction, which builds unified joint graphs via syntactic dependency trees, LLM-generated triples, and entity linking; Graph Optimization, employing the Graph Information Bottleneck (GIB) to prune redundant nodes/edges and enhance cross-modal interactions; and Multimodal Fusion, where concatenated structural and semantic features are utilized to predict entity relations.

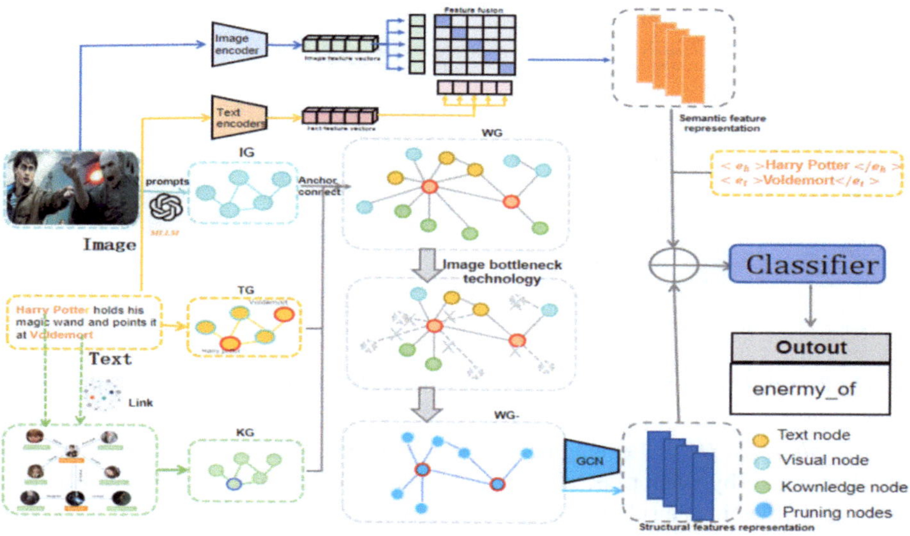

Fig. 1. The overview of our model structure

3.2 Semantic Representation

BERT model transforms textual input into normalized sequences, yielding dense vector representations. The Vision Transformer (ViT) partitions images into non-overlapping patches, producing visual tokens. Cross-modal fusion facilitates the alignment of modalities within a shared latent space. Bidirectional attention mechanisms enhance feature representations, and the unified representation is derived through the integration of these features $\widehat{v}$.

$$\widehat{V} = W_v V + b_v,\ \widehat{T} = W_t T + b_t \tag{1}$$

$$H_t = Softmax\left(\frac{\widehat{T}\widehat{V}^T}{\sqrt{d}}\right)\widehat{V},\ H_v = Softmax\left(\frac{\widehat{V}\widehat{T}^T}{\sqrt{d}}\right)\widehat{T} \tag{2}$$

$$\hat{v} = ReLU(W_f[H_t \| H_v] + b_f) \tag{3}$$

3.3 Cross - Modal Joint Graph

Subgraph Construction

Text Subgraph (TG): Constructed from syntactic dependency trees using ELMo-based parsing (Peters et al. 2018), nodes represent words, and edges encode grammatical dependencies (e.g., subject-verb-object relations).

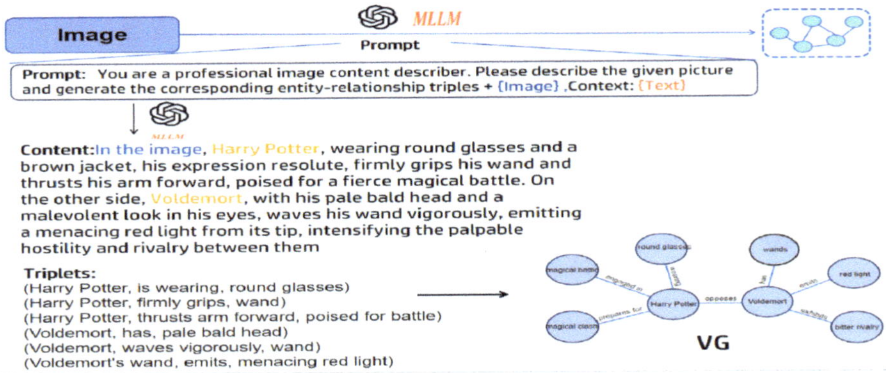

Fig. 2. Utilize large models to generate descriptive triples, thereby constructing the IG.

Image Subgraph (IG): Leveraging Large Language Models (LLMs), visual features V extracted by a vision encoder are fused with textual prompts to generate structured triples *(h,r,t)*.The image subgraph(Fig. 2) $IG = (V_1, E_1)$.

Knowledge Graph (KG): Entity linking integrates semantic similarity Sim(e)) and structural centrality Centrality(e)Centrality(e).

Knowledge - Text Anchor Alignment

The joint graph $WG = (V,E)$ integrates TG,IG, KG subgraphs via anchor alignment. Cross-modal edges. (E^{TK}, E^{TI}) between anchors are weighted by attention:

$$w_{i,j} = Softmax(Att(h_i, h_j)) \tag{4}$$

where h_i, h_j are node embeddings.

3.4 Optimizing the Graph Structure

Each For each node v_k and edge $e_{k,l}$, learnable parameters μ_k^v,$\mu_{k,l}^e$ determine their retention through differentiable Bernoulli sampling (Jang et al. 2017). Contextual features (m-order neighbors) and target entities $(h_s; h_o)$ are aggregated via attention:

$$\mathrm{s_k^v} = \mathrm{Att}(\mathrm{v_k}, \phi(v_k); \mathrm{H}), \;\; s_{k,l}^e = Att(v_k, \phi(v_k), v_l, \phi(v_l); H) \tag{5}$$

$$\mu_{k}^{v} = \mathrm{Sigmoid}(\mathrm{FFN}([s_{k}^{v}; h_{s}; h_{o}])),\ \mu_{k,l}^{e} = Sigmoid(FFN([s_{k,l}^{e}; h_{s}; h_{o}])) \quad (6)$$

The GCN propagates features using normalized adjacency $\widehat{A} = \widehat{A} = \widehat{D}^{-\frac{1}{2}}\widehat{A}\widehat{D}^{-\frac{1}{2}}$. Yielding graph representation $\hat{y}$.

$$H^{(l+1)} = ReLU(\widehat{A}H^{(l)}W^{(l)}) \quad (7)$$

$$\hat{y} = \frac{1}{N}\sum\nolimits_{i=1}^{N} H_{i}^{(L)} \quad (8)$$

3.5 Relation Prediction

We fuse structural feature $\hat{y}$, semantic feature $\hat{v}$, and context from head-tail entity pair(e_h, e_t) to form $\hat{z} = Concat(\hat{v}, \hat{y}, e_h, e_t)$. This is then input into an MLP followed by a softmax for label probability prediction. we using the cross-entropy loss function.

$$\mathrm{p}(\mathrm{r}|\hat{z}) = \mathrm{softmax}(\mathrm{MLP}(\hat{z})) \quad (9)$$

$$\mathrm{L} = -\sum\nolimits_{i=1}^{n} \log(\mathrm{p}(\mathrm{r}|\hat{z})) \quad (10)$$

4 Experiments

4.1 Experimental Setup

The MNRE benchmark comprises 15,485 aligned text-image pairs (23 relations, 30,970 entities) with 15% weakly aligned pairs for robustness evaluation. Multimodal graphs are constructed via: (1) Text Graph: BERT-base dependency parsing; (2) Image Graph: GPT-3.5-generated structured triplets; (3) Knowledge Graph: FAISS-aligned Wikidata entries with dynamic scoring. The unified graph is pruned via GIB ($\tau = 0.1$, $\epsilon \sim U(0,1)$), removing 35–41% noise, then refined by 2-layer GAT (8 heads) and 3-layer GCN ([768,256,128]).

Training uses AdamW (lr = 2e−5, batch = 32) for 50 epochs with dropout (p = 0.3) and label smoothing ($\alpha = 0.1$), validated on a 20% holdout set (patience = 5). Evaluation spans accuracy, F1, speed, and 20% synthetic noise robustness. Baselines include text-only (BERT, DP-GCN), text-image (ViLBERT, MEGA), and ablation variants.

4.2 Experimental Results

Table 1. Performance Comparison on MNRE Dataset

Method	Accuracy (%)	Precision (%)	Recall (%)	F1 (%)
Text-Only				
PCNN (2015)	73.15	62.85	49.69	55.49
MTB (2019)	75.69	64.46	57.81	60.86
PURE (2021)	73.03	64.50	58.19	61.22
Text-Image				
VisualBERT (2019)	–	57.15	59.48	58.30
MEGA (2021)	80.05	64.51	68.44	66.41
GPT4-XMLR (2023)	–	–	–	74.56
EEGA (2023)	–	78.27	78.91	78.59
DGF-PT (2023)	–	79.72	78.63	79.24
Text-Image-KG				
KITE-MRE(Ours)	**82.73**	**82.00**	**77.54**	**79.71**

KITE-MRE achieves state-of-the-art performance on MNRE (82.73% accuracy, 79.71% F1; Table 1), outperforming text-only baselines (PCNN, MTB, PURE; <62% F1) constrained by semantic ambiguity and multimodal gaps, as well as recent multimodal methods (MEGA, VisualBERT: ~66% F1; EEGA, DGF-PT: ~78–79% F1) limited by static fusion and rigid knowledge. The framework addresses these via dynamic KG fusion (adaptively weighting triples by multimodal coherence) and graph refinement (pruning spurious edges), achieving optimal precision (82.00%) while maintaining recall (77.54%). Hierarchical pruning and adaptive attention further reduce computational overhead for real-time deployment. Remaining challenges in long-tail relations and unseen entities motivate adaptive KG augmentation.

4.3 Ablation Study

Ablation experiments on MNRE validate core component necessity (Table 2). Removing the Text Graph (TG) causes the largest F1 drop (−6.30%), notably impairing recall (−7.84%) by losing syntactic distinctions (e.g., "CEO_of" vs. "advisor_to"). Image Graph (IG) ablation reduces recall (−4.97%), failing to disambiguate spatially grounded relations (e.g., defaulting to "located_near" instead of "tourist_photographing"). Knowledge Graph (KG) removal degrades long-tail relation F1 (−5.2%, e.g., misclassifying "patent_by" as "develop"). Disabling GIB optimization marginally lowers F1 (−2.25%) but amplifies noise sensitivity (73.46% vs. 79.71% F1 under 20% noise).

Table 2. Ablation Study Results

Model Variant	Accuracy (%)	Precision (%)	Recall (%)	F1 (%)
Full Model	**82.73**	**82.00**	**77.54**	**79.71**
- Text Graph	77.34 (−5.39)	77.33 (−4.67)	69.70 (−7.84)	73.41 (−6.30)
- Image Graph	79.66 (−3.07)	79.58 (−2.42)	72.57 (−4.97)	75.99 (−3.72)
- Knowledge Graph	80.22 (−2.51)	79.77 (−2.23)	73.76 (−3.78)	76.69 (−3.02)
- GIB Optimization	81.43 (−1.30)	80.61 (−1.39)	74.49 (−3.05)	77.46 (−2.25)

The full model synergizes TG (syntactic disambiguation), IG (visual grounding), KG (semantic anchoring), and GIB (noise suppression). Persistent challenges include handling KG-absent entities (7.1% F1 drop for emerging brands) and temporal relations ("succeed_to": 69.8% F1), necessitating future work on expandable KGs and temporal-aware modeling.

5 Conclusion

KITE-MRE, a knowledge-infused multimodal relation extraction framework, addresses semantic fragmentation and auxiliary information gaps by integrating LLM-generated image triplets and KG-entity alignment into a unified graph. Dynamic pruning via Graph Information Bottleneck (GIB) eliminates redundancies, while GCN-based cross-modal fusion enhances relation prediction. Evaluated on MNRE, it achieves state-of-the-art performance (82.73% accuracy, 79.71% F1), resolving KG incompleteness and LLM ambiguity. Ablation studies confirm critical contributions of text/image graphs and KG integration, with robustness to noise via multimodal fusion. Future directions include adaptive KG augmentation for emerging entities and temporal-aware representations for dynamic scenarios.

References

Bollacker, K., Evans, C., Paritosh, P., Sturge, T., Taylor, J.: Freebase: a collaboratively created graph database for structuring human knowledge. In: Proceedings of the 2008 ACM SIGMOD international conference on Management of data (2008)

Lu, J., Batra, D., Parikh, D., Lee, S.: ViLBERT: pretraining task-agnostic visiolinguistic representations for vision-and-language tasks. In: Proceedings of the IEEE/CVF International Conference on Computer Vision (ICCV), pp. 13–23. IEEE (2019)

Chen, Y.-C., et al.: Uniter: Universal image-text representation learning. In European Conference on Computer Vision, pp. 104–120. Springer (2020)

Chen, H., Huang, Y., Takamura, H., Nakayama, H.: Commonsense knowledge aware concept selection for diverse and informative visual storytelling. In Proceedings of the AAAI Conference on Artificial Intelligence, vol. 35 (2021)

Chen, X., et al.: Hybrid transformer with multi-level fusion for multimodal knowledge graph completion. In: Proceedings of the 45th International ACM SIGIR Conference on Research and Development in Information Retrieval (2022)

Chen, X., Zhang, N., Li, L., Yao, Y.: Good visual guidance makes a better extractor: hierarchical visual prefix for multimodal entity and relation extraction

Dong, J., Zhang, Q., Zhou, H., Zha, D., Zheng, P., Huang, X.: Modality-aware integration with large language models for knowledge-based visual question answering. arXiv preprint arXiv: 2402.12728 (2024)

Dosovitskiy, A.: An image is worth 16 × 16 words: transformers for image recognition at scale. arXiv preprint arXiv:2010.11929 (2020)

Feng, J., et al.: Towards bridged vision and language: learning cross-modal knowledge representation for relation extraction. IEEE Trans. Circuits Syst. Video Technol. **34**(1), 561–575 (2023)

Xu, H., et al.: Pre-trained models: past, present and future. AI Open **2**, 225–250 (2021)

Hu, X., Chen, J., Liu, A., Meng, S., Wen, L.: Prompt me up: Unleashing the power of alignments for multimodal entity and relation extraction. In: Proceedings of the 31st ACM International Conference on Multimedia, pp. 5185–5194 (2023)

Buschoff, L.M.S., Akata, E., Bethge, M., et al.: Visual cognition in multimodal large language models. Nat. Mach. Intell. **7**(1) (2025). https://doi.org/10.1038/s42256-024-00963-y

Chen, Y.C., Li, L., Yu, L., et al.: UNITER: UNiversal Image-TExt Representation Learning

Zheng, C., Feng, J., Fu, Z., et al.: Multimodal relation extraction with efficient graph alignment. In: Proceedings of the 29th ACM International Conference on Multimedia

Zheng, C.: Multimodal relation extraction with efficient graph alignment. In: Proceedings of the 29th ACM International Conference on Multimedia (MM '21), pp. 5298–5306. Association for Computing Machinery, New York, NY, USA (2021)

A Lightweight Data Indexing Mechanism with Low Query and Maintenance Overhead

Jingyu Zhang[1(✉)], Di Lan[1], Zisang Xu[1], and Jin Wang[2]

[1] School of Computer Science and Technology, Changsha University of Science and Technology, Changsha 410076, China
zhangzhang@csust.edu.cn, randy233@stu.csust.edu.cn

[2] Sanya Institute of Hunan University of Science and Technology, Sanya 572024, China
jinwang@hnust.edu.cn

Abstract. Edge storage systems aim to store data on edge nodes to provide users with reliable and low-latency data retrieval services. This is especially important for latency-sensitive applications, where users need to locate the target data and retrieve it quickly under multi-hop delay constraints. Although users can query neighboring nodes through serial or parallel retrieval methods, these approaches often lead to significant redundant communication and resource consumption across the edge network. Therefore, designing an efficient indexing mechanism is essential. However, current research on indexing mechanisms for local retrieval in edge storage systems remains limited. In this paper, we propose a tree-based indexing structure built upon the probabilistic data structure known as the Counting Bloom Filter (CBF). By incorporating the advantages of Delaunay Triangulation (DT) graph, the proposed indexing structure enables efficient data query and retrieval. Experimental results show that, compared with existing local retrieval indexing structures such as EDIndex-CBF and EDIndex-HCBF, our approach achieves lower maintenance and query overhead.

Keywords: Edge Computing · Data Query · Distributed Retrieval

1 Introduction

In recent years, the rapid proliferation of smart devices has driven an exponential increase in data traffic. It is projected that by 2029, the number of mobile devices worldwide will reach approximately 90 billion, with an annual data growth rate of up to 26% [1, 2]. Relying solely on cloud platforms to process this vast amount of data can lead to issues such as bandwidth pressure, increased latency, and privacy risks—problems that are particularly critical in scenarios like autonomous driving [3], federated learning [4], and smart healthcare [5]. In contrast, edge computing [6] mitigates these challenges by bringing computation and storage closer to end users, making it a prominent focus of current research.

Data indexing mechanisms [7–9] are one of the key technologies for achieving efficient data sharing in edge systems. The core of this lies in improving the query and

J. S. Park et al. (Eds.): FutureTech 2025, LNEE 1483, pp. 165–171, 2025.
https://doi.org/10.1007/978-981-95-1999-6_21

retrieval efficiency of data within the edge network. By constructing a rational data indexing structure, the system can quickly locate the target data's position, thereby reducing access latency and enhancing overall response speed. Figure 1 shows a typical scenario of efficient data sharing.

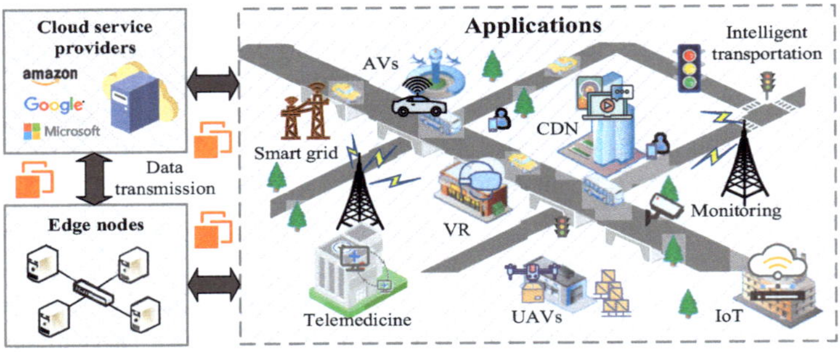

Fig. 1. Application scenarios of edge data sharing.

Current data indexing mechanisms mainly include global indexing and local indexing. Global indexing, such as the distributed hash table (DHT) in a structured peer-to-peer network [10, 11], distributes data to nodes with unique identifiers through hash functions to achieve efficient retrieval. Coordinate-based indexing (COIN) [12] uses node coordinates to build a forwarding table to achieve low-latency data location. Although global indexing has high search efficiency, it ignores access patterns and may cause high latency. At the same time, the local indexing mechanism is still imperfect, which motivates the research of this paper.

This paper proposes a lightweight query and maintenance overhead (LQMO) indexing structure, based on Count Bloom Filters (CBF) [13]. LQMO adopts a tree structure composed of two types of CBFs. CBF^{ord} records node data, while CBF^{aggre} records aggregated data from neighbors. In our designed experiments, compared with two other local index structures, the proposed index structure has lower query cost and maintenance overhead.

2 LQMO Indexing Mechanism

2.1 Design of LQMO

The LQMO indexing mechanism shares similarities with EDIndex-CBF [14]. As shown in Fig. 2, each node in the LQMO index consists of a CBF, used to record data items of a single node (CBF^{ord}) or an aggregation of multiple node (CBF^{aggre}). Specifically, $CBF_{i,j}^{aggre}$ represents the aggregated data of v_i's j-hop neighbors.

To better illustrate the design and advantages of the LQMO indexing, we compare it with EDIndex-CBF using a local Delaunay Triangulation (DT) graph under a two-hop delay constraint. Taking node v_1 as an example (Fig. 2a), its EDIndex-CBF structure is shown in Fig. 2b. The root stores v_1's local data, while the first level stores CBF^{ord} for its 1-hop neighbors v_2 to v_6, and the second level stores CBFs for 2-hop neighbors v_7 and v_8. Since EDIndex-CBF maintains each neighbor in an individual manner, a query at worst requires traversing the entire index tree, resulting in higher overhead.

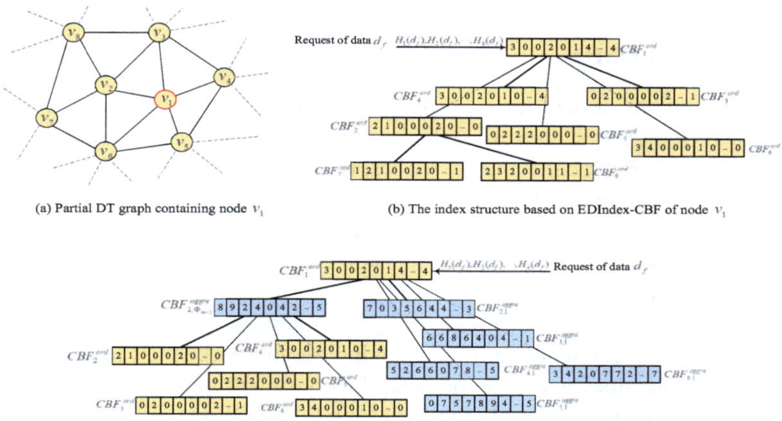

(a) Partial DT graph containing node v_1 (b) The index structure based on EDIndex-CBF of node v_1

(c) The LQMO indexing structure of node v_1

Fig. 2. Indexing structure of EDIndex-CBF and LQMO.

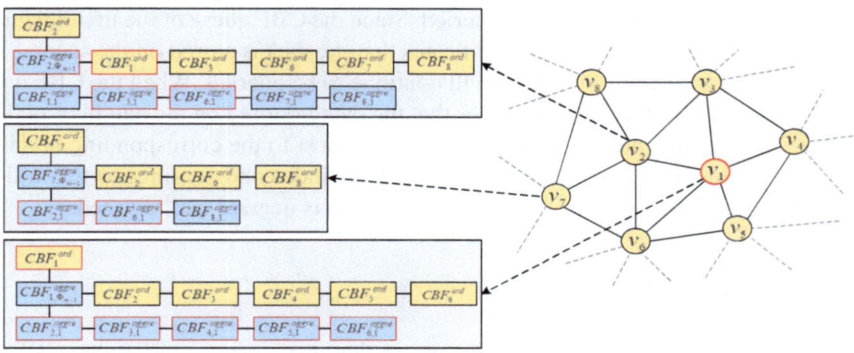

Fig. 3. Specific implementation of LQMO.

In contrast, the LQMO (Fig. 2c) retains the local CBF^{ord} at the root but introduces a more efficient subtree design. The first subtree includes CBF^{ord} s for all neighbors within $\Phi_m - 1$ hops (in this case, 1-hop), and its root node contains a CBF^{aggre} aggregating their data to accelerate queries. When data is not found in earlier hops, this aggregation speeds up queries.

The remaining subtrees (each consisting of a single node) store aggregated data of the 1-hop neighbors of the $\Phi_m - 1$-th hop neighbor. These "single-node subtrees" ensure that queries and retrievals can be efficiently forwarded to the Φ_m-th hop node even when the index maintains $\Phi_m - 1$ hop information.

The LQMO is implemented using a unified linked-list structure (Fig. 3), with three lists: the first stores v_1's local CBF^{ord}; the second stores the first subtree, with the head node containing aggregated data of its children; and the third stores aggregated information of each $\Phi_m - 1$-hop neighbor's 1-hop neighborhood.

2.2 Querying in LQMO

When a node initiates a data retrieval request, the system first checks its local index structure to see if the target data exists. For example, for node v_1 with a delay constraint of 2, if the data is stored locally and the request is initiated by node v_1, the query process will be completed in the first linked list of the local index. Since this list contains only one list item, the query directly checks the CBF of this list item. If the CBF returns a successful query, the data exists locally, and node v_1 can read the data from local storage, ending the retrieval process.

When the data is stored at one-hop neighbor nodes of node v_1 and the request is initiated by v_1, the query will begin by checking the first linked list in the local index. Since the local node does not store the data, the query will then proceed to the second linked list. Until a certain linked list item is found, its CBF judgment data is stored in the node, then node v_1 will forward the request to the corresponding neighbor node. Assuming the data is stored at neighbor v_2, the request will be forwarded to v_2, and v_2 will return the data.

When the data is stored on the two-hop neighbor node of v_1 and the query is initiated by v_1, the process first queries the first linked list. At this time, the query fails and the second linked list will continue to be queried. Since the CBF query of the first linked list item in the second linked list fails, it indicates that the data is stored on the second-hop neighbor node, so the third linked list will continue to be queried. When the CBF query of a linked list item succeeds, it indicates that the data is stored on the neighbor node of the node. At this time, node v_1 will forward the request to the corresponding neighbor node (for example, v_2). Then, v_2 will check its local index and continue to forward the request to its neighbor node, and so on, until the data is queried and returned.

3 Experimental Evaluation

We compare the proposed method with state-of-the-art approaches EDIndex-CBF and EDIndex-HCBF. The experimental parameters are configured as follows: the number of nodes varies from 100 to 350, increasing by 10 in each step, and the delay constraint varies between 1-hop, 2-hop, and 3-hop.

3.1 Maintenance Overhead of LQMO

This study compares the maintenance overhead of LQMO, EDIndex-CBF, and EDIndex-HCBF, as shown in Fig. 4. Under a 1-hop delay constraint Φ_m, EDIndex-HCBF has the highest overhead. To reduce lookup latency, EDIndex-HCBF must maintain two additional CBF^{aggre}: one for the 1-hop neighbors and another for the combined aggregation of all CBF^{ord}. In contrast, LQMO and EDIndex-CBF have lower overhead, as they don't

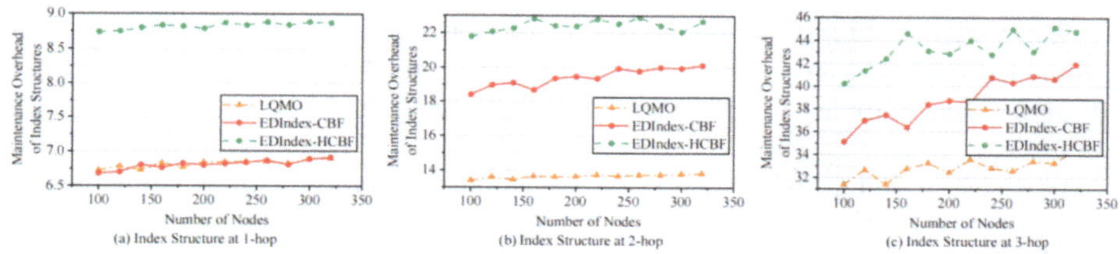

Fig. 4. Index maintenance overhead with different hops.

require additional CBFs at this stage, and their index structures are almost identical, resulting in similar costs.

Under a 2-hop delay constraint, EDIndex-HCBF needs to maintain three extra CBF^{aggre}: one for 1-hop neighbors, one for 2-hop neighbors, and one for the combined all data aggregation. This leads to the highest maintenance overhead. EDIndex-CBF, however, does not need additional CBF^{aggre}, making its overhead lower than EDIndex-HCBF's. LQMO, compared to EDIndex-CBF, maintains fewer CBFs under this delay constraint. EDIndex-CBF maintains CBFs for all neighbors within the 2-hop range plus the local node, while LQMO only needs to maintain the 1-hop neighbors' CBF^{ord} s and an additional CBF^{aggre} s for the aggregation of each 1-hop neighbor's neighbors. Although this increases LQMO's CBF maintenance, its total overhead is still lower than EDIndex-CBF.

Similarly, under a 3-hop delay constraint, EDIndex-CBF must maintain 3-hop neighbors, while LQMO only needs to maintain 2-hop neighbors and the related additional CBFs. Experimental results further confirm the advantages of the LQMO indexing structure.

3.2 Query Overhead of LQMO

This part compares the data query overhead of LQMO with the index structures based on Index-CBF and Index-HCBF under different delay constraints. As shown in Fig. 5, when the delay constraint Φ_m is set to 1, EDIndex-HCBF incurs the highest query overhead. Although it maintains two additional CBF^{aggre} s to speed up queries, one is used to store the aggregated data of one-hop neighbors. This is because when the data is stored in a one-hop neighbor or a local node, the data existence check of the CBF^{aggre} s will be judged successful, resulting in additional query overhead.

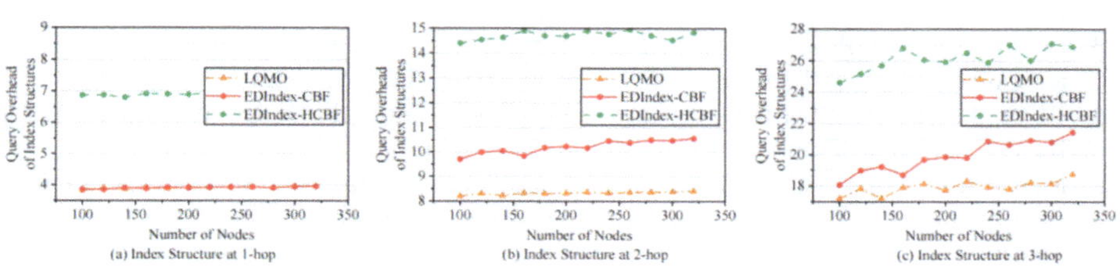

Fig. 5. Index query overhead with different hops.

When Φ_m= 2, LQMO introduces some additional query overhead due to the extra maintained CBF. However, its overall query cost is still lower than that of EDIndex-HCBF because EDIndex-HCBF maintains more CBF^{ord} and CBF^{aggre} structures than LQMO.

When Φ_m= 3, LQMO achieves the lowest query overhead, while EDIndex-CBF exhibits the highest. This is because under this delay constraint, the number of 3-hop neighbors maintained by EDIndex-CBF increases significantly, reducing the likelihood of successful queries from 1-hop or 2-hop neighbors. EDIndex-HCBF mitigates unnecessary query delay by leveraging additional CBF^{aggre} structures. Meanwhile, LQMO not only reduces unnecessary query delays through extra CBF^{aggre}, but can also skip the lookup of 1-hop and 2-hop neighbors if the data is stored on a 3-hop neighbor, by directly locating it via the CBF^{aggre}.

4 Conclusion

This paper proposes a lightweight data indexing mechanism LQMO, which contains two CBFs: CBF^{ord} s and CBF^{aggre} s. One is used to record the data storage information of the node, and the other is used to record the data aggregation information of neighboring nodes. We design the experiments to verify the effectiveness of proposed LQMO method. Compared with state-of-the-art schemes,

LQMO outperforms the other methods in terms of query and maintenance overhead in different testing scenarios.

Acknowledgement. This research is supported by the Hunan Provincial Key Research and Development Program (No. 2024JK2005), the National Natural Science Foundation of China (No. 62402064), the Natural Science Foundation of Hunan Province, China (Nos. 2023JJ30054, 2024JJ6066), the Hunan Provincial Natural Science Foundation Key Program (No. 2024JJ3017), and the Research Foundation of Education Bureau of Hunan Province, China (Nos. 23B0303, 23B0288).

References

1. Ericsson: Ericsson mobility report. https://www.ericsson.com/4912e3/assets/local/reports-papers/mobility-report/documents/2024br/emr-business-review-2024.pdf
2. Seagate, IDC: Rethink data: Put more of your business data to work—from edge to cloud. https://www.seagate.com/our-story/rethink-data/RethinkDataReport2020.pdf
3. Li, M., Gao, J., Zhao, L., Shen, X.: Adaptive computing scheduling for edge-assisted autonomous driving. IEEE Trans. Veh. Technol. **70**(6), 5318–5331 (2021)
4. Wang, S., et al.: Adaptive federated learning in resource constrained edge computing systems. IEEE J. Sel. Areas Commun. **37**(6), 1205–1221 (2019)
5. Meher, B.K., Amin, R., Abdussami, M., Sureshkumar, V., Hossain, M.A.: Efficient certificateless anonymous mutual authentication in wbans for smart healthcare. IEEE Trans. Intell. Transp. Syst. **25**(11), 17666–17675 (2024)
6. Shi, W., Cao, J., Zhang, Q., Li, Y., Xu., L.: Edge computing: vision and challenges. IEEE Internet Things J. **3**(5), 637–646 (2016)

7. Monga, S.M.,Sheshadri, K.R., Simmhan, Y.: ElfStore: a resilient data storage service for federated edge and fog resources. ICWS 2019, pp. 336–345
8. Guo, D., Xie, J., Shi, X., Cai, H., Qian, C., Chen, H.: HDS: a fast hybrid data location service for hierarchical mobile edge computing. IEEE/ACM Trans. Netw. **29**(3), 1308–1320 (2021)
9. Xie, J., Qian, C., Guo, D., Li, X., Wang, G., Chen, H.: A novel data placement and retrieval service for cooperative edge clouds. IEEE Trans. Cloud Comput. **11**(1), 71–84 (2023)
10. Stoica, I., Morris, R., Karger, D.R., Kaashoek, M.F., Balakrishnan, H.: Chord: a scalable peer-to-peer lookup service for internet applications. SIGCOMM 2001, pp. 149–160
11. Augustine, J., Chatterjee, S., Pandurangan, G.: A fully-distributed scalable peer-to-peer protocol for byzantine-resilient distributed hash tables. SPAA 2022, pp. 87–98
12. Xie, J., Qian, C., Guo, D., Wang, M., Ge., Wang, Chen, H.: COIN: an efficient indexing mechanism for unstructured data sharing systems. IEEE/ACM Trans. Netw. **30**(1), 313–326 (2022)
13. Ficara, D., Giordano, S., Procissi, G., Vitucci, F.: MultiLayer compressed counting bloom filters. INFOCOM 2008, pp. 311–315
14. He, Q., et al.: EDIndex: enabling fast data queries in edge storage systems. SIGIR 2023, pp. 675–685

A Novel PSO-Based Adaptive Container Scheduling Strategy for Edge Computing

Jingyu Zhang[1(✉)], Hanbo Jiang[1], Zisang Xu[1], and Jin Wang[2]

[1] School of Computer Science and Technology, Changsha University of Science and Technology, Changsha 410076, China
{zhangzhang,xzsszx111}@csust.edu.cn, jianghb@stu.csust.edu.cn
[2] Sanya Institute of Hunan University of Science and Technology, Sanya 572024, China
jinwang@hnust.edu.cn

Abstract. Edge computing has emerged as a critical technology for achieving low latency and real-time data processing, especially in applications with the Internet of Things (IoTs) and smart cities. With their lightweight and portable nature, containers provide an effective solution for deploying and managing edge applications. The Particle Swarm Optimization (PSO) algorithm, inspired by the collective behavior of bird flocks, offers an efficient approach to complex optimization problems by iteratively updating particle positions within a multidimensional solution space. PSO's low computational complexity and strong global search capability make it highly suitable for resource allocation and task scheduling in edge environments. This paper presents an adaptive scheduling strategy based on PSO. In this approach, a time-series-based Dynamic Request Prediction Algorithm (DRPA) anticipates request patterns to dynamically adjust container placement, minimizing response time. Concurrently, a threshold-based PSO (TPSO) optimizes container scheduling based on node load, enhancing resource utilization and reducing scheduling delays. Experimental results show that DRPA-TPSO effectively lowers latency, boosts resource utilization, and improves system stability in high-load, complex environments, outperforming traditional scheduling methods.

Keywords: Particle Swarm Optimization · Edge Computing · Container Scheduling · Adaptive Scheduling · Dynamic Threshold

1 Introduction

Edge computing, driven by 5G and IoT, reduces latency by placing servers near data sources. Edge containers enhance this by integrating container technology, ensuring low latency, scalability, and efficiency. Figure 1 show the typical scenario of edge container architecture. However, PSO, commonly used for optimization, struggles with heterogeneous edge nodes and dynamic task demands, limiting its scheduling effectiveness. Edge computing optimizes task allocation for performance, latency, and resource efficiency. Sonmez et al. [1] developed EdgeCloudSim, while Zhou et al. [2] improved scheduling with prioritization and reinforcement learning. Despite progress, scalability

J. S. Park et al. (Eds.): FutureTech 2025, LNEE 1483, pp. 172–178, 2025.
https://doi.org/10.1007/978-981-95-1999-6_22

and dynamic adaptability issues persist, requiring further research for low-latency, high-efficiency task scheduling. Container virtualization enables applications to run in isolated environments while sharing the OS kernel. Zhang et al. [3] optimized task scheduling, while Alves et al. [4] proposed a lightweight container scheme for resource-constrained edge devices. Hu et al. [5] applied ML for workload prediction to optimize resource allocation. Kubernetes, as a mainstream orchestration platform [6], faces scalability and scheduling bottlenecks. Despite advancements, Kubernetes still faces inefficiencies in scheduling, network bottlenecks, and security challenges, requiring further optimization for real-time edge applications. Particle Swarm Optimization (PSO), a heuristic algorithm, is widely used in edge computing for its efficiency [7]. Wang et al. [8] optimized earthquake emergency supply scheduling with an enhanced PSO model.

Despite advancements, PSO still faces challenges in stability and adaptability to dynamic edge environments. To tackle this, we propose DRPA-TPSO, combining Dynamic Request Prediction (DRPA) for adaptive container placement and Threshold-PSO (TPSO) for optimized edge node allocation. DRPA analyzes user behavior for predictive scheduling, while TPSO adjusts node loads dynamically, improving resource utilization and reducing transmission overhead. Our key contributions include:

1. We propose the DRPA algorithm, which predicts future service requests using historical data and real-time trends to optimize container placement and task execution.
2. We also propose the TPSO strategy, which dynamically adjusts container scheduling based on predicted requests, reducing failure rates in resource-intensive tasks and improving efficiency.
3. DRPA-TPSO is proposed, and we experimentally tested it in terms of resource allocation, task execution, and system stability. The experimental statistics show the effectiveness of our proposed method.

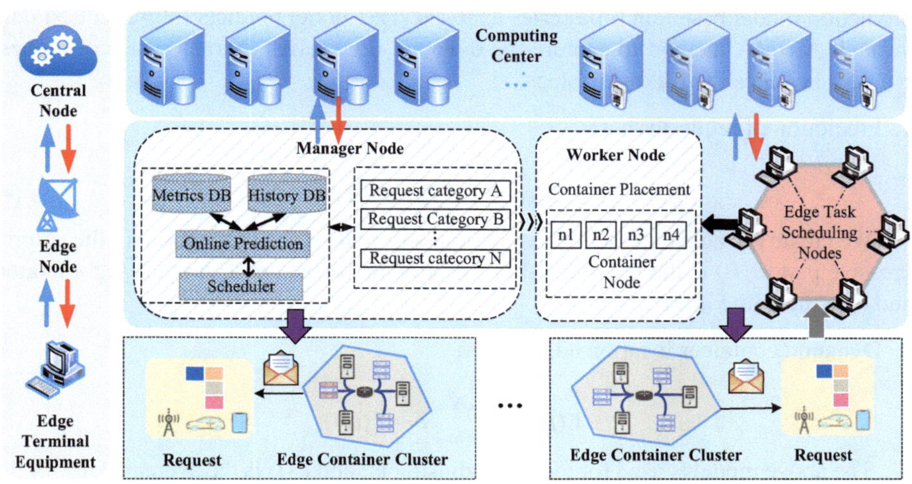

Fig. 1. The typical scenario of edge container architecture.

2 Design of DRPA-TPSO

We propose DRPA-TPSO (see Fig. 2 for the overall system framework) to adaptively minimize response time as well as maximize resource utilization. First, the time series-based request type prediction algorithm DRPA utilizes historical data to predict future request types. By training time series analysis tool ARIMA (Autoregressive Integral Sliding Average) model, the system is able to predict the changes of request types and thus adjust the location of containers in advance to minimize the task response time.

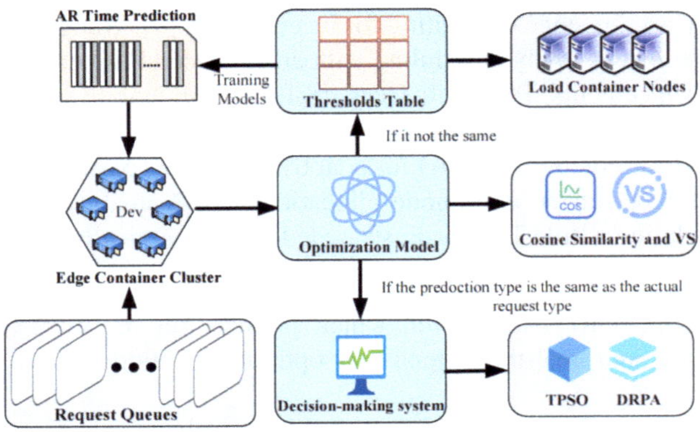

Fig. 2. Overall system framework.

3 DRPA Prediction Models

Predicting changes in device request types in edge computing environments is crucial to reduce the system response time. To achieve this goal, this paper introduces a request type prediction model based on time series analysis. The model predicts future request data through ARIMA method to capture the dynamic time series characteristics of request types. It includes the following steps:

1. Prediction of request types:

$$\phi(B)(1 - B)^d y_t = \theta(B)\epsilon_t \tag{1}$$

We create the above model for prediction, and $\phi(B)$ is the polynomial in the autoregressive part, $\theta(B)$ is the polynomial in the moving average part, B is the lag operator, and ϵ_t is the white noise.

2. Dynamic container location adjustment:

$$\mathrm{L}(j) = \sum\nolimits_{i=1}^{N} p_i \cdot d(i, j) \tag{2}$$

The above model is used for location adjustment, and L(j) is the average delay of the container at node j, p_i is the probability of request i, $d(i, j)$ is the delay between request i and node j.

4 TPSO Algorithm

In order to optimize the container adaptive scheduling strategy in the edge computing environment, this paper proposes a Threshold-Based PSO algorithm (TPSO) based on a threshold table. The algorithm dynamically establishes and adjusts the threshold table by analyzing the load of each edge node, which is used to guide the scheduling and allocation of computing tasks. The specific methods are as follows:

1. TPSO create threshold table based on the overall workload, which can be expressed as follows:

$$\mathrm{Load}_i = \sum_{i=1}^{N} (w_{\mathrm{cpu}} \times \frac{U_{\mathrm{cpu,i}}}{T_{\mathrm{cpu}}} + w_{\mathrm{mem}} \times \frac{U_{\mathrm{mem,i}}}{T_{\mathrm{mem}}} + w_{\mathrm{net}} \times \frac{U_{\mathrm{net,i}}}{T_{\mathrm{net}}}) \quad (3)$$

   where w is the weight of different resources, U is the CPU, memory, and network bandwidth usage of the i node, and T is the total threshold of the corresponding resource.
2. Threshold Setting: the TPSO system sets specific thresholds to determine when to start, stop or migrate containers. For example, system shall automatically starts new containers when their CPU utilization exceeds a certain threshold, and automatically stops containers when their CPU utilization falls below a certain threshold.
3. Task Scheduling Optimization: the tpso algorithm fully considers the dynamics and complexity in the edge computing environment. The particle position update formula is:

$$\mathbf{x}_i(t+1) = \mathbf{x}_i(t) + \mathbf{v}_i(t+1) \quad (4)$$

   where $\mathbf{x}_i(t)$ is the current position of the i particle in generation t and $\mathbf{v}_i(t+1)$ is the velocity of the particle.The particle velocity update formula is:

$$\mathbf{v}_i(t+1) = w\mathbf{v}_i(t) + c_1 r_1(\mathbf{p}_i - \mathbf{x}_i(t)) + c_2 r_2(\mathbf{g} - \mathbf{x}_i(t)) + \alpha \cdot \Delta T_i \quad (5)$$

   where α is the coefficient associated with the threshold table and ΔT_i is the difference between the threshold table values corresponding to the current position $\mathbf{x}_i(t)$ of the particle, calculated as:

$$\Delta T_i = \mathrm{Threshold}(\mathbf{x}_i(t)) - U_{\mathrm{current}}(\mathbf{x}_i(t)) \quad (6)$$

   here, $\mathrm{Threshold}(\mathbf{x}_i(t))$ is the threshold at the corresponding position in the threshold table and $U_{\mathrm{current}}(\mathbf{x}_i(t))$ is the load at the current position. In conjunction with the threshold table, the fitness function can be set to minimize delay or maximize resource utilization while meeting the threshold requirements:

$$f(\mathbf{x}_i) = min(\mathrm{L}(\mathbf{x}_i) + \lambda \cdot \mathrm{OverThreshold}(\mathbf{x}_i)) \quad (7)$$

where OverThreshold($\mathbf{x}_i$) is the amount of load that exceeds the threshold table at the current position $\mathbf{x}_i$ and λ is the penalty factor. Dynamically adjust the target position of the particle based on the comparison of the current particle position with the threshold table:

$$\mathbf{g} = \arg\min_{\mathbf{x}_i}(\text{OverThreshold}(\mathbf{x}_i)) \tag{8}$$

here, $\mathbf{g}$ is the global best position in the particle swarm. The particle swarm adaptive optimization algorithm based on threshold table fully considers the dynamics and complexity in the edge computing environment.

5 Optimization Models of DRPA-TPSO

1. Judgment on Predicted Type. DRPA-TPSO shall judge whether the predicted type is the same as the actual type. If the calculated cosine similarity is close to 1, it means that the actual request is highly similar to the predicted request types, which can be used for subsequent operations. If the cosine similarity is low, it means that the prediction results do not match the actual situation. In this case, the request needs to be added into the prediction model, and the prediction model needs to be re-evaluated. $\mathbf{R} = [R_1, ..., R_i]$ denotes the actual request type of the edge device, $\mathbf{P} = [P_1, ..., P_i]$ denotes the predicted request type, and their cosine similarity is as follows:

$$Consine\ Similarity = \frac{\sum_{i=1}^{n} R_i \times P_i}{\sqrt{\sum_{i=1}^{n} R_i^2} \times \sqrt{\sum_{i=1}^{n} P_i^2}} \tag{9}$$

2. Task Scheduling. If predicted request type matches, the decision-making system is activated to quickly assign it to the corresponding node through the DRPA-TPSO algorithm, according to the following scoring model.

$$Score = \alpha \cdot \frac{T_{expected}}{T_{threshold}} + \beta \cdot \frac{C_{request}}{C_{threshold}} + \gamma \cdot QoS_{score} \tag{10}$$

where α, β, γ denote the weights of response time, computational complexity and QoS demand respectively. If Score is less than the synthesis threshold $S_{threshold}$, the request is judged to be a low latency request, otherwise it is a high performance request.

6 Experimental Evaluation

The experiments are conducted on an NFS Server (see Table I for the specific configuration) using Kuburnete to manage the containerized cluster, which consists of two management nodes and six worker nodes. The NFS server is installed on a Macbook machine, and all the data in this experiment is stored on the NFS server. The EdgeCloudSim simulation platform is employed, and we use a real AWS Lambda request log dataset with it. We compare the key metrics, including average response time and resource utilization.

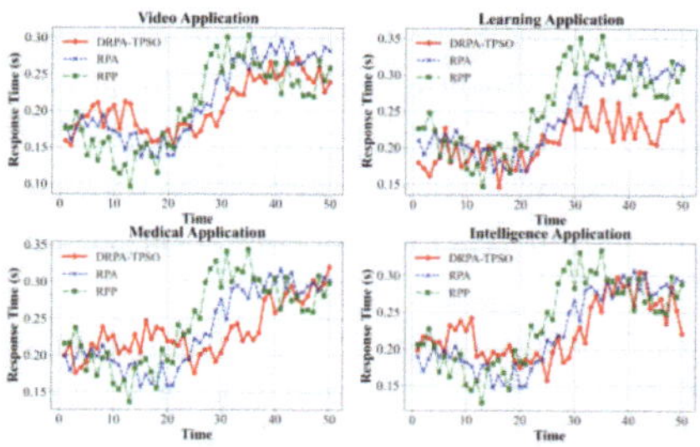

Fig. 3. Comparisons on the response time.

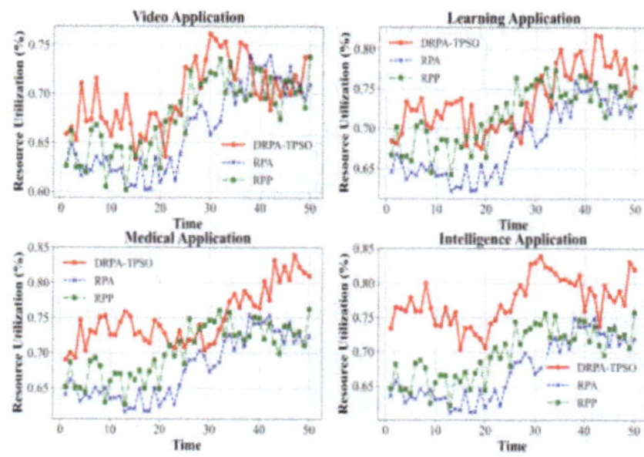

Fig. 4. Comparisons on the resource utilization.

We compare the performance of the task scheduling algorithm DRPA-TPSO with the following two benchmarks: **1. RPA:** its fullname is Resource Partitioned Allocation, and its core idea is to allocate system resources to different tasks or nodes based on demand, ensuring stability and efficiency when handling high loads. Each node is treated equally in this method. **2. RPP:** its fullname is Resource Provisioning based on Prediction, which dynamically allocates resources based on demand and load predictions, ensuring optimized and stable system performance through real-time adjustments and efficient resource utilization.

As shown in Fig. 3, the experimental results show that, compared with the benchmark algorithms RPA and RPP, DRPA-TPSO achieved significantly lower average response time across all four application scenarios. This is because DRPA-TPSO can effectively reduce task processing time, according to appropriate prediction and threshold setting, thereby enhancing system real-time performance and user experience.

The resource utilization comparisons are shown in Fig. 4, and it shows DRPA-TPSO also demonstrated superior performance in resource utilization. The experimental statistics indicate that DRPA-TPSO achieved higher resource utilization in all application scenarios compared with RPA and RPP. The reason is, DRPA-TPSO can allocate the different tasks with different needs to suitable low-load nodes, through the adaptive global optimization mechanism.

7 Conclusion

Edge computing, as the critical technology, can provide the low latency and real-time processing capacity for IoT or Smart City application. With their lightweight and portable nature, containers provide an effective solution for deploying and managing applications in the edge computing environments. This paper presents a novel approach called DRPA-TPSO to enhance task scheduling and resource allocation efficiency in containerized edge computing environments. The method optimizes task scheduling and resource utilization through a two-stage process. In the first stage, DRPA-TPSO uses a time-series prediction model to forecast the request type and quantity of service requests, dynamically adjusting container placement based on historical data analysis to reduce latency and improve system responsiveness. In the second stage, DRPA-TPSO combines

POS algorithm with a threshold-based adaptation mechanism to dynamically adjust container scheduling based on real-time node load, maximizing resource utilization and reducing scheduling delay. The designed experiments showed the proposed DRPA-TPSO outperforms state-of-the-art methods, in terms of response time and resource utilization. In the future, we plan to utilize DRPA-TPSO in the real-world heterogeneous scenarios.

Acknowledgment. This research is supported by the Hunan Provincial Key Research and Development Program (No. 2024JK2005), the National Natural Science Foundation of China (No. 62402064), the Natural Science Foundation of Hunan Province, China (Nos. 2023JJ30054, 2024JJ6066), the Hunan Provincial Natural Science Foundation Key Program (No. 2024JJ3017), and the Research Foundation of Education Bureau of Hunan Province, China (Nos. 23B0303, 23B0288).

References

1. Mahmud, M.R., Toosi, A.N.: Con-Pi: a distributed container-based edge and fog computing framework. IEEE Internet Things J. **9**(6), 4125–4138 (2022)
2. Zhou, W., et al.: Priority-aware resource scheduling for UAV-mounted mobile edge computing networks. IEEE Trans. Veh. Technol. **72**(7), 9682–9687 (2023)
3. Zhang, J., Zhou, X., Ge, T., Wang, X., Hwang, T.: Joint task scheduling and containerizing for efficient edge computing. IEEE Trans. Parallel Distrib. Syst. **32**(8), 2086–2100 (2021)
4. Alves, M.P., Delicato, F.C., dos Santos, I.L., Pires. P.F.: LW-CoEdge: a lightweight virtualization model and collaboration process for edge computing. World Wide Web **23**(2), 1127–1175 (2020)
5. Hu., Shihong, Shi, W., Li, G.: CEC: a containerized edge computing framework for dynamic resource provisioning. IEEE Trans. Mob. Comput. **22**(7), 3840–3854 (2023)
6. Pitonak, R., Mucha, J., Dobis, L., Javorka, M., Marusin, M.: CloudSatNet-1: FPGA-based hardware-accelerated quantized CNN for satellite on-board cloud coverage classification. Remote. Sens. **14**(13), 3180 (2022)
7. Yu, Y., Yin, S.: A comparison between generic algorithm and particle swarm optimization. ISAIMS 2020, pp. 137–139
8. Dong, S., Xia, Y., Kamruzzaman, J.: Quantum particle swarm optimization for task offloading in mobile edge computing. IEEE Trans. Ind. Inform. **19**(8), 9113–9122 (2023)

A Tree-Based Pipeline Consensus with High Scalability and Throughput

Jingyu Zhang[1(✉)], Sheng Jiang[1], Zisang Xu[1], and Jin Wang[2]

[1] School of Computer Science and Technology, Changsha University of Science and Technology, Changsha 410076, China
{zhangzhang,xzsszx111}@csust.edu.cn, sjiang@stu.csust.edu.cn
[2] Sanya Institute of Hunan University of Science and Technology, Sanya 572024, China
jinwang@hnust.edu.cn

Abstract. Edge computing has become an increasingly popular paradigm as part of distributed computing architecture. It accomplishes this by facilitating data from end devices to be stored and processed at the edge of the network close to the data. Unfortunately, consensus scalability and throughput become a big issue in collaborative edge scenarios. Byzantine consensus is a promising consensus solution that is feasible in small-scale collaborative edge systems. However, most of the existing consensus algorithms cannot meet the increasing number of nodes in collaborative edges, which poses a big challenge to the throughput of consensus algorithms. In this paper, we propose a tree-based pipeline consensus, a variant BFT (Byzantine Fault Tolerant) that maintains high throughput as the system size grows, utilizing a novel pipelining technique that performs scalable splitting and recombination on the tree to increase the efficiency of BLS (Boneh-Lynn-Shacham). We conduct experiments through simulations. The experimental results show that the proposed consensus mechanism can effectively improve the performance of collaborative edges, including throughput and consensus latency.

Keywords: Edge Computing · Blockchain · Byzantine Fault Tolerant

1 Introduction

With the rapid development of Internet of Things (IoT) technology, cloud computing is increasingly unable to meet the needs of terminals due to network latency and bandwidth limitations [1, 2]. In edge computing, data owned by stakeholders is rarely shared with each other due to privacy issues and the huge cost of data transmission. Therefore, the opportunities for cooperation among multiple stakeholders are limited. To overcome this weakness, collaborative edge solutions have been designed for edge devices [3], which connect the edges of multiple stakeholders distributed geographically regardless of their physical location and network structure [4]. When edge devices perform collaborative services, fault tolerant consensus is a crucial part of edge computing [5]. With fault tolerant consensus, edge devices can reach effective consensus on task offloading, final decision-making, etc.

J. S. Park et al. (Eds.): FutureTech 2025, LNEE 1483, pp. 179–185, 2025.
https://doi.org/10.1007/978-981-95-1999-6_23

However, the number of nodes required in edge computing environments is increasing, and the emerging use cases of permissioned blockchains require the system participants to expand into the hundreds [6]. Most permissioned blockchains are based on variants of the Byzantine Fault Tolerant (BFT) consensus protocol [7]. These protocols have poor scalability and can no longer meet the node requirements of today's systems.

To address the above issues, we propose tree-based pipeline consensus (TBPC), a BFT communication concept that leverages the split/recombination tree (S/R tree) for load balancing and scalability, avoiding the poor throughput of previous tree-based solutions due to the latency of additional consensus rounds. Our proposed TBPC adopts a novel tree topology, placing the leader at the root node and distributing the load among all internal nodes of the tree. In addition, the TBPC adopts multi-signature BLS (Boneh-Lynn-Shacham) [8] to reduce the message size, which is complementary to the characteristics of the TBPC, reducing the message propagation consumption while also reducing the leader's load.

In summary, this paper makes the following contributions: (1) we design models to support the use of S/R tree in the context of consensus protocols; (2) we propose the TBPC, an efficient communication consensus for the BFT protocol, which achieves higher throughput than HotStuff in multiple scenarios and better latency under certain conditions; (3) experimental results confirm the superiority of the proposed TBPC when the node size of the system increases.

2 Tree-Based Pipeline Consensus

2.1 Framework of S/R Tree

Based on the current traditional communication mode, HotStuff is an algorithm that uses a star topology to achieve consensus. Based on HotStuff, this paper proposes a new method, which replaces the star structure of HotStuff with the topology of an S/R tree. We analyze the communication mode used in HotStuff and discuss it in detail.

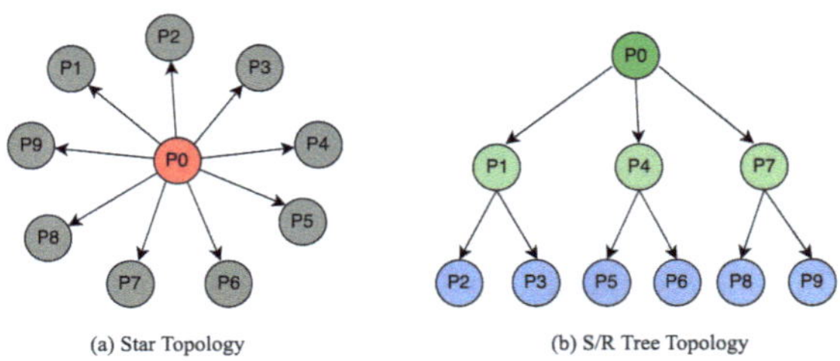

Fig. 1. HotStuff and S/R tree communication patterns.

Each round consists of two phases: (i) the split phase, in which the leader splits some information to all nodes; (ii) the recombination phase, in which the leader aggregates and recombines messages from other nodes. All rounds use a unified mode, but the information split and recombined by the leader in each round is different. Figure 1(a) illustrates the communication pattern of HotStuff in a system with 9 processes.

We now discuss how to implement the split and recombination phases using a tree topology while maintaining the same properties. This process is illustrated in Fig. 1(b). The following is an introduction to the S/R tree.

Data Splitting. This algorithm is used in the first phase of each round to split the transaction from the leader node to all other child nodes. The child nodes then forward the data to their own child nodes, and so on.

Vote Recombination. This algorithm is used in the second phase of each round so that the leader recombines *N-f* signatures for arbitration. The parent node then recombines these signatures with its own signature and sends the recombined signature to its own parent node. This process is repeated until the final recombined signature is calculated at the root of the tree.

2.2 Encrypted Collection

The TBPC adopts a multi-signature BLS scheme, allowing each internal node to aggregate the votes of its child nodes into an aggregate vote. The load complexity of each internal node (including the root node) is $O(m)$, where m is the fanout of the tree, and the complexity of verifying the aggregate vote is $O(1)$. In the case of a large scale of collaborative edge nodes, multiple signatures need to be integrated and verified, and BLS is more suitable.

The core advantage of the BLS threshold signature algorithm lies in its efficiency and security. When processing a large number of signatures, the BLS algorithm can provide faster verification speeds than traditional digital signature algorithms, which is particularly important for distributed systems and blockchain applications that need to frequently verify signatures. In addition, since the verification of aggregate signatures does not depend on specific signers, even if an attacker obtains the private keys of some signers, it is impossible to forge a valid aggregate signature, which provides additional security for the system.

2.3 Tree-Shaped Pipeline Structure

As a traditional communication mode, the star topology has certain advantages, but it cannot meet the system throughput when the node scale is large, and its scalability is not good. The S/R tree we proposed in this paper solves the scalability problem, but the latency is usually higher than that of the star topology, so we propose a special tree-based pipeline structure.

In the tree-based pipeline, we take full advantage of the flexible, robust, and load-balanced properties of the k-any tree. In the k-any tree, the fanout m is much smaller than the number of nodes N, so in the tree-based pipeline, the root node completes its propagation phase much faster than in HotStuff, and it may be idle before it starts collecting votes. This allows the root node to start multiple consensus instances during the execution of a single consensus round. This introduces a multiplicative factor, Pipelength, which increases the pipeline depth of HotStuff. The increase in pipeline depth allows higher parallelism and thus improves throughput (Fig. 2).

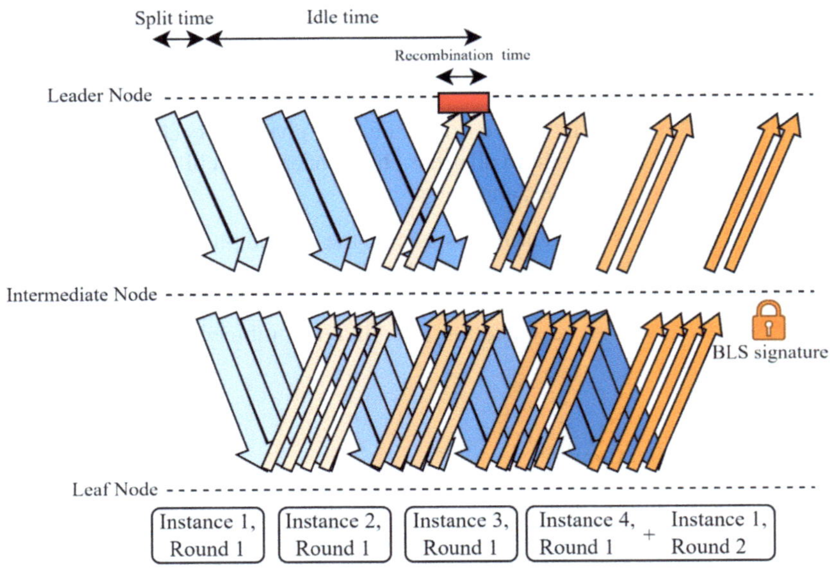

Fig. 2. Tree-based pipeline structure.

Pipelength, the number of instances that can be started in a round, is affected by the following parameters:

Split Time: the time it takes for a node to send a block to all its children. This value is a function of the fanout *m*, the block size *B*, and the link bandwidth *b*, and is approximately *mB/b*.

Processing Time: the time it takes a node to verify and/or aggregate the votes received from its children. This depends heavily on the type of signatures used by the algorithm. We shall experimentally measure this for different signature schemes.

Idle Time: the time elapsed from when the root node finished sending a block to its children until the last reply was received and processed. This value is the sum of the network latency and the processing time defined above. It is roughly given by:

$$Idle\ time \ = \ h \cdot (RTT \ + \ processing\ time) \tag{1}$$

where *h* is the height of the S/R tree and *RTT* is the network round trip time. In a star topology, the idle time is short and is mainly used to collect signatures. However, in an S/R tree, the root node is usually idle for most of the running time. Tree-based pipeline takes advantage of this large idle time to launch additional consensus instances. Therefore, the challenge is to estimate how many additional instances can be launched, i.e., Pipelength.

The Pipelength of the tree-based pipeline allows us to take full advantage of the time saved by the leader and only need to interact with *m* nodes instead of *N-1* nodes. Therefore, the ratio between *N-1* and *m* defines the maximum speedup ratio that can be achieved using an S/R tree structure instead of a star structure.

3 Evaluation

The experimental environment runs on a host with Microsoft Windows 10, 16 GB RAM, 1TB SSD, and AMD Ryzen 7 3600X processor. The host provides 100 Mbps bandwidth, 6 physical cores, 12 threads, and 128 GB memory. In this section, we evaluate the TBPC in several scenarios. Each measurement is the average of 5 runs, and the error bar represents one standard deviation. We experimentally vary these baseline parameters to illustrate their impact on performance.

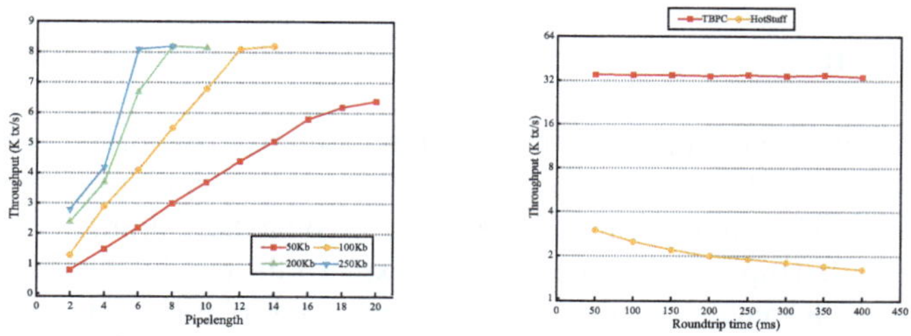

(a) Effect of Pipelength on TBPC throughput

(b) Impact of RTT in system throughput

Fig. 3. The pipelengh and RTT comparisons.

3.1 The Impact of Pipelength on Tree-Based Pipeline Throughput

Figure 3(a) shows the effect of pipelength on the throughput of the proposed TBPC (Tree-Based Pipeline Consensus) for $N = 100$, $m = 10$, and different block sizes. As can be seen in Fig. 3(a), for $250Kb$ block size, the best results are obtained when Pipelength is close to 6. And, when the block size is smaller, a higher Pipelength is required to fully utilize resources.

Pipelining not only allows TBPC to better utilize available computing and network resources but also helps to mitigate the negative impact of the additional RTT inherent in tree-based methods. Figure 3(b) indicates the impact of RTT in system throughput. When N = 100 and the bandwidth is $100Mb/s$, the throughput of HotStuff-secp decreases as the RTT increases, while the throughput of TBPC can be kept almost unchanged by increasing the Pipelength to avoid the leader being idle while waiting for a response.

3.2 Throughput Comparisons of Different Consensus Algorithms

We compare the throughput of TBPC with HotStuff-secp and HotStuff-bls with different scales of nodes. To evaluate the impact of our pipeline scheme, we also include the results of TBPC without pipelines (denoted as TBPC-np) by using a system with BLS signatures but without pipeline-based trees. The results are shown in Fig. 4. The experimental results show that TBPC outperforms other methods on the total network throughput.

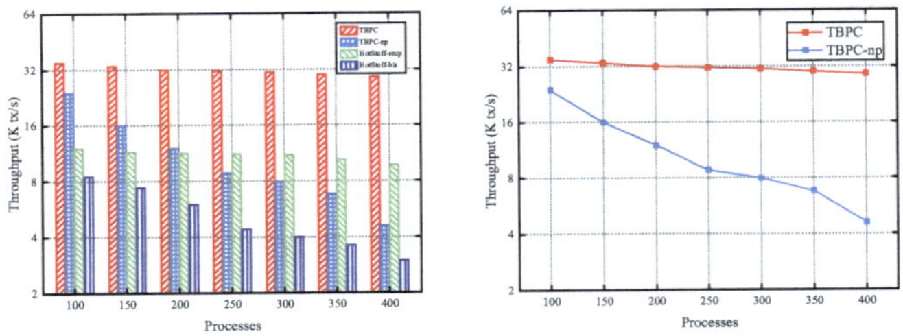

Fig. 4. Throughput comparisons for different configurations.

4 Conclusion

Current state-of-the-art permissioned blockchains suffer from important limitations: bottlenecks caused by work concentration on the leader, reduced throughput when load distribution is enforced, or reduced resilience. We overcome these limitations by introducing a new tree-based pipeline scheme that fully exploits the parallelization opportunities provided by the S/R tree. In contrast to committee-based solutions, our proposed TBPC does not compromise the resilience or finality of consensus. Experiments show that the throughput of TBPC is significantly better than HotStuff in all considered cases, with only a slight increase in latency. TBPC also outperforms HotStuff in terms of throughput and latency when bandwidth is constrained.

Acknowledgment. This research is supported by the Hunan Provincial Key Research and Development Program (No. 2024JK2005), the National Natural Science Foundation of China (No. 62402064), the Natural Science Foundation of Hunan Province, China (Nos. 2023JJ30054, 2024JJ6066), the Hunan Provincial Natural Science Foundation Key Program (No. 2024JJ3017), and the Research Foundation of Education Bureau of Hunan Province, China (Nos. 23B0303, 23B0288).

References

1. Simon, W.A., Qureshi, Y.M., Rios, M., Levisse, A., Zapater, M., Atienza, D.: BLADE: an in-Cache computing architecture for edge devices. IEEE Trans. Comput. **69**(9), 1349–1363 (2020)
2. Ardakani, A., Condo, C., Gross, W.J.: Fast and efficient convolutional accelerator for edge computing. IEEE Trans. Comput. **69**(1), 138–152 (2020)
3. Yu, W., et al.: A survey on the edge computing for the internet of things. IEEE Access **6**, 6900–6919 (2018)
4. Shi, W., Cao, J., Zhang, Q., Li, Y., Xu, L.: Edge computing: vision and challenges. IEEE Internet Things J. **3**(5), 637–646 (2016)
5. Jing, G., Zou, Y., Yu, D., Luo, C., Cheng, X.: Efficient fault-tolerant consensus for collaborative services in edge computing. IEEE Trans. Computers **72**(8), 2139–2150 (2023)
6. Kokoris-Kogias, E., Jovanovic, P., Gasser, L., Gailly, N., Syta, E., Ford, B.: OmniLedger: a secure, scale-out, decentralized ledger via sharding. In: IEEE Symposium on Security and Privacy, pp. 583–598 (2018)

7. Lamport, L., Shostak, R., Pease, M.: The Byzantine generals problem. ACM Trans. Program. Lang. Syst. **4**(3), 382–401 (1982)
8. Boneh, D., Lynn, B., Shacham, H.: Short signatures from the weil pairing. J. Cryptol. **17**(4), 297–319 (2004)

Design and Evaluation of P-GAN for Privacy Protection and Generative AI Regulatory Compliance

Seo Jeong Min[1], Ji Su Park[2], and Jin Gon Shon[1](✉)

[1] Department of Computer Science, Korea National Open University, Seoul, South Korea
{honyangi89,jgshon}@knou.ac.kr

[2] Department of Computer Science and Engineering, Jeonju University, Jeonju, South Korea
jisupark@jj.ac.kr

Abstract. This study deals with the design and implementation of a learning model to prevent the leakage of sensitive information of the original data during the learning process of a model that generates high-quality data that is difficult to distinguish from the original using a generative adversarial network (GAN). Existing European General Personal Information Protection Act (GDPR) and domestic Personal Information Protection Act had limitations in that it was difficult to effectively respond to the privacy leakage occurring in the characteristics of the Generative AI model and the learning method. In order to improve this problem, AI Act that applies differential regulations for each risk level of the system was introduced. Therefore, this study aims to introduce personal information enhancement technology (PETs) into the GAN model design process to promote a balance between the usefulness of the learning model and privacy enhancement, and to propose a new P-GAN model that can quantitatively evaluate the learning results. In this paper, by describing the design principles and implementation results of the proposed model in detail, we seek the direction of improvement of the future generative model's usefulness and privacy protection function enhancement system.

Keywords: GAN · P-GAN · AI Act · PETs

1 Introduction

This study addresses the design and implementation of a learning model intended to prevent the leakage of sensitive information from original data during the training of Generative Adversarial Networks (GANs) [1–3], which are employed to generate high-quality synthetic data ostensibly indistinguishable from the source. The authors contend that existing regulatory frameworks, such as the European General Data Protection Regulation (GDPR) and domestic South Korean privacy laws, possess inherent limitations in effectively mitigating privacy risks arising from the specific characteristics and training methodologies of generative AI models [4, 5]. It is noted that AI legislation, applying differential regulations based on system risk levels, has been introduced to ameliorate these identified shortcomings.

J. S. Park et al. (Eds.): FutureTech 2025, LNEE 1483, pp. 186–193, 2025.
https://doi.org/10.1007/978-981-95-1999-6_24

Consequently, the research proposes a novel P-GAN model architecture. This involves the integration of Privacy Enhancing Technologies (PETs) [2, 6, 7] within the GAN design process, purportedly to establish a balance between model utility and enhanced privacy safeguards. A further claim is that this proposed model facilitates the quantitative evaluation of learning outcomes. The paper aims to provide a detailed exposition of the proposed model's design principles and implementation results, ostensibly to inform future directions for improving systems that enhance the utility and privacy-preserving capabilities of generative models.

2 Privacy Regulations and GAN Design Requirements

The GAN model aiming to comply with personal information protection regulations must consider both the technical requirements and the measures necessary for regulatory compliance. Specifically, these considerations can be classified into a review of privacy protection technologies, vulnerability analysis, and adherence to regulatory standards. Differential privacy [2, 8, 9] is a mathematical model that safeguards personal information by adjusting the intensity of privacy applied to the learning model, thereby minimizing the influence of individual records in the training data on the overall dataset.

Differentially Private Stochastic Gradient Descent (DPSGD) [2, 9] is a privacy protection technique that combines clipping, which controls gradient L2 norm calculated by individual data points, and Gaussian noise injection mechanisms to limit the sensitivity of individual data points to model training. The P-GAN proposed in this study utilizes the Tensorflow_privacy library to compute privacy losses and track the history of occurrences simultaneously with the calculation of privacy losses, thereby calculating the privacy losses and tracking the history of loss occurrences. Throughout the learning process, RDP implementation is implemented to verify budget overruns in real time through dynamic adjustment of noise injection, and efficiently manage the losses accumulated through this information, thereby achieving an optimal trade-off between the model's utility and privacy protection.

Federated Learning (FL) [10, 11] enables training across distributed environments without centralized storage of original data, thereby reducing exposure risks. This approach proves particularly valuable in sectors where privacy safeguards are paramount, including financial institutions and public administration. Despite these advantages, multiple vulnerability vectors persist—notably membership inference attacks [12, 13] that could potentially compromise source data confidentiality. To address these challenges while bolstering model resilience, researchers are investigating comprehensive protection frameworks incorporating several complementary strategies: adversarial regularization techniques, privacy parameter optimization, normalization protocols, and strategic early termination of training to mitigate overfitting risks [2, 7, 12].

Creating AI architectures that successfully balance privacy preservation with data utility while meeting regulatory requirements demands enhanced security protocols aligned with protection standards outlined in GDPR and applicable privacy legislation. A growing consensus acknowledges that conventional regulatory frameworks alone provide insufficient safeguards against sophisticated threats emerging during AI-driven data generation—particularly re-identification vulnerabilities and inference-based information extraction [6, 13, 14]. This recognition catalyzed the European Union's introduction of the Artificial Intelligence Act (AI Act) in 2020. The legislation implements a tiered regulatory approach, classifying systems according to their inherent risk profiles and applying proportionate oversight mechanisms to each category. Consequently, AI systems developed henceforth must demonstrate verifiable compliance with these newly established criteria. Ensuring trustworthy model evaluation outcomes therefore necessitates implementing diverse risk indicators to quantitatively assess regulatory adherence levels [6].

In light of this, Designing artificial intelligence models necessitates achieving equilibrium between personal information protection and data utility while simultaneously adhering to regulatory standards. This imperative calls for strengthened security measures meticulously aligned with the data protection policies stipulated in GDPR and analogous privacy legislation. Nevertheless, a broadening realization exists that conventional legal statutes, including GDPR, struggle to adequately regulate the complex risks—such as re-identification vulnerabilities and information leakage through inference [6, 13, 14]—that arise during AI systems' data generation processes. This evolving regulatory landscape prompted the European Union to introduce its Artificial Intelligence Act (AI Act) policy framework in 2020. Fundamentally, the AI Act aims to implement differentiated regulations based on a classification of AI systems according to their potential risk levels.

3 P-GAN: Design of GAN Models Compliant with Personal Data Protection Regulations

3.1 P-GAN Architecture with Integrated Privacy-Enhancing Technologies (PETs)

Foundational Design Principles of the P-GAN Model Sharing Framework.

○ Safeguarding Sensitive Information: The framework incorporates advanced Privacy-Enhancing Technologies (PETs) [2, 6, 7] to establish comprehensive encryption protocols and internal data protection mechanisms, effectively preventing sensitive information exposure within learned model parameters. This methodological approach fundamentally eliminates potential data breach vectors during model training phases and prevents reconstruction of embedded personal identifiers through reverse engineering techniques [13, 14]. Distinctive to this implementation is the application of robust protection protocols across both generator and discriminator components—the architectural cornerstones of GAN models [1, 3]—ensuring integrity throughout the entire learning pipeline.

○ Regulatory Alignment: The mechanism addresses technical limitations inherent in existing GDPR frameworks and personal data protection legislation while incorporating tiered risk assessment protocols as outlined in the AI Act. This forward-looking strategy transcends baseline compliance requirements to provide adaptive resilience against evolving regulatory landscapes. By implementing security measures proportionate to assessed risk levels in AI development contexts, the framework minimizes unnecessary constraints on innovation while establishing enhanced protective barriers for high-risk application domains.

○ Utility-Security Equilibrium: The design implements computationally efficient encryption algorithms and streamlined governance structures to ensure that enhanced security measures maintain operational harmony with the model's generative capabilities and data utility objectives. This balanced approach preserves analytical value while reinforcing protective safeguards.

3.2 P-GAN Model Sharing Mechanism

In this section, we introduce a novel mechanism for securely sharing P-GAN models. This approach leverages differential privacy principles [2, 8, 9, 15] enhanced with Gaussian noise to enable the safe distribution of GAN models that generate high-quality synthetic data without revealing sensitive information [16–18]. By integrating the latest regulatory shifts and privacy protection requirements, the mechanism is designed to preclude data leakage during training while maintaining an optimal balance between model utility and privacy. The secure sharing protocol consists of the following procedures:

○ Gradient Clipping: During each training batch, the L2 norm of the discriminator gradients is restricted to a predefined threshold. This constraint, often part of techniques like DPSGD [2, 9], minimizes the impact of individual data points on model updates, thereby reducing the model's sensitivity when processing specific training samples.

○ Noise Injection: To further obscure individual data contributions, Gaussian noise is added to the clipped gradients, a common practice in differentially private training [2, 9]. The noise's standard deviation is dynamically determined by hyper parameters—such as the clipping threshold, noise multiplier, and batch size—that are tailored to the operational environment. This stochastic masking effectively prevents the extraction of identifiable details from the gradients.

○ Privacy Budget Administration: A differential privacy mechanism [2, 8, 9] is employed to quantitatively measure the cumulative privacy loss incurred throughout the training process. Expressed in terms of a privacy loss boundary (ε) and a failure probability (δ), these parameters are calculated with tools like the Tensorflow_privacy library. Monitoring the accumulated ε enables dynamic adjustments to the number of training epochs and noise levels, ensuring that the overall privacy expenditure remains within the preset budget.

○ Intrinsic Data Protection: By incorporating Privacy-Enhancing Technologies (PETs) [2, 6, 7] directly within the P-GAN model parameters, the dependency on individual training samples is statistically minimized. This built-in safeguard effectively constrains any sensitive information from being ingrained in the final model representation.

○ Quantitative Privacy Assurance: The final, shared P-GAN model is provided along with the computed terminal ε and δ values. These quantitative metrics serve as verifiable evidence of the level of privacy protection achieved during training [cf. 6], offering transparent insights to model recipients regarding compliance with strict privacy standards.

4 Performance and Regulatory Compliance Assessment of P-GAN Models

This evaluation comprehensively assesses the P-GAN model's compliance with AI and data protection regulatory requirements using the Flickr-Faces-HQ (FFHQ) dataset [3]—a high-resolution facial image collection containing potentially sensitive information. Given the inherent sensitivity of facial data, robust privacy protection mechanisms and regulatory compliance are essential for ensuring both the reliability and social acceptability of the P-GAN model. The training process was executed through a minimum of 300 iterations to ensure stability, with early termination criteria established at 150 iterations to maintain computational efficiency.

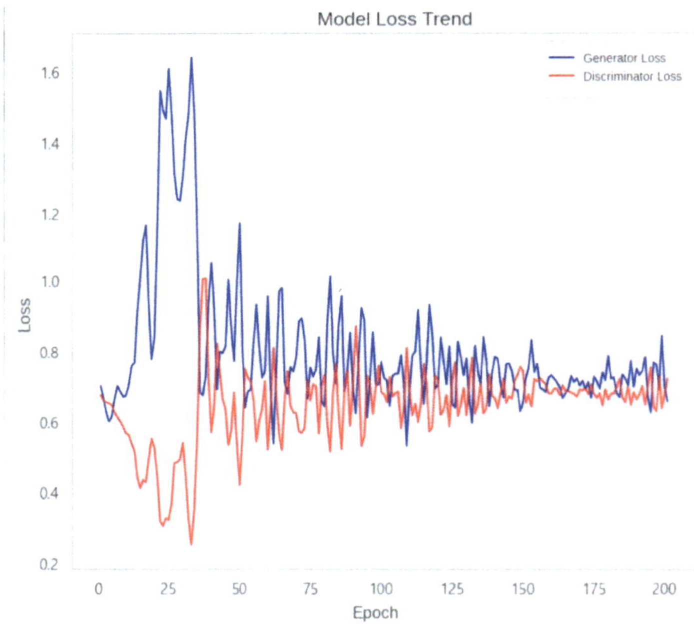

Fig. 1. P-GAN Model Loss Trend

The assessment yielded the following key findings:

- AI Act Core Requirements Fulfillment: Quantitative evaluation across multiple dimensions considered fundamental in the AI Act including technical robustness, data governance, and transparency demonstrated that P-GAN consistently exceeded the established compliance thresholds across all evaluation metrics. These results provide objective evidence of the model's capacity to satisfy diverse regulatory requirements specified in current AI legislation.
- Verification of Embedded Privacy Protection Design: Fig. 1 confirms the effective implementation of differential privacy mechanisms during model training [2, 6, 7],

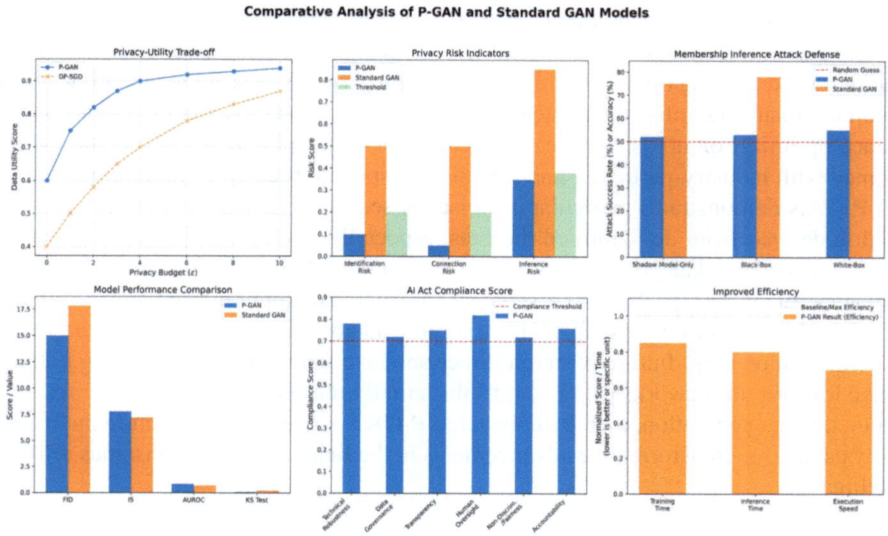

Fig. 2. Dashboard for Regulatory Compliance Evaluation

demonstrating systematic management of privacy budget consumption (cumulative ε). The successful application of these techniques with the FFHQ dataset [3], which presents significant re-identification risk [4, 12], validates the incorporation of essential technical measures required by the "Privacy by Design" principle mandated in the GDPR and related regulations. Furthermore, this approach establishes a framework for quantitative assessment and management of privacy protection levels specified in regulatory frameworks [4].

- Evidence of Data Protection Principle Adherence: As illustrated in Fig. 2, P-GAN demonstrates significantly reduced quantified privacy risk levels [4] and exhibits robust defense against membership inference attacks (MIAs) [10, 11] designed to extract individual data points. Additionally, Kolmogorov-Smirnov testing, which measures distribution similarity between generated and original data, indicates that P-GAN maintains lower statistical divergence compared to control models, suggesting superior preservation of statistical properties without compromising privacy. These findings collectively demonstrate that P-GAN effectively safeguards privacy by preventing exposure of specific data within the FFHQ dataset [3] while maintaining utility in data-driven applications, thereby addressing critical requirements in data protection and security regulations.

5 Conclusion

This study set out to evaluate whether P-GAN models can effectively balance privacy protection with data utility while meeting regulatory requirements. The evaluation dashboard findings empirically substantiate P-GAN's exceptional performance across multiple dimensions: privacy preservation, data utility, attack resilience, model efficacy, regulatory compliance, and computational efficiency. This systematic analysis enables precise quantification of P-GAN's conformity with regulatory frameworks such as GDPR and the AI Act during operational data training scenarios. While acknowledging certain computational trade-offs, the strategic balance between security protocols and performance optimization holds significant potential for developing domain-specific metrics aligned with industry-particular and jurisdiction-specific regulatory requirements.

P-GAN demonstrates particular promise for secure management of sensitive information derived from de-identified datasets, especially within healthcare and financial ecosystems where data privacy concerns are paramount [8, 9, 14]. The model's ability to withstand membership inference attacks [10, 11] while maintaining statistical similarity to original data distributions represents a significant advancement in privacy-preserving generative modeling. Future research directions involving privacy-preserving collaborative learning frameworks [8, 9] could substantially mitigate inference vulnerability in synthetic data generation, thereby enhancing P-GAN's security architecture and potentially catalyzing transformative advancements in the field of privacy-conscious artificial intelligence.

References

1. Arjovsky, M., Chintala, S., Bottou, L.: Wasserstein GAN. arXiv preprint arXiv:1701.07875 (2017)
2. Fan, L.: A survey of differentially private generative adversarial networks. arXiv preprint arXiv:1911.01727 (2019)
3. Karras, T., Laine, S., Aila, T.: A style-based generator architecture for generative adversarial networks. In: Proceedings of the IEEE/CVF Conference on Computer Vision and Pattern Recognition, pp. 4401–4410. IEEE, Los Alamitos (2019)
4. Goodman, B., Flaxman, S.: European Union regulations on algorithmic decision-making and a "right to explanation." AI Mag. **38**(3), 50–57 (2017)
5. Ko, H., Park, S.: How to de-identify personal data in South Korea: an evolutionary tale. Int. Data Priv. Law **10**(4), 385–394 (2020)
6. Giomi, M., Boenisch, F., Wehmeyer, C., Tasnádi, B.: A unified framework for quantifying privacy risk in synthetic data. Proc. Privacy Enhan. Technolog. **2023**(2), 5–25 (2023)
7. Zhu, J., Xia, Y., Sun, S., Li, G., Wu, F.: Privacy-preserving adversarial network. In: Advances in Neural Information Processing Systems, vol. 32, pp. 1–12. Curran Associates, Inc. (2019)
8. Jordon, J., Yoon, J., Van Der Schaar, M.: PATE-GAN: generating synthetic data with differential privacy guarantees. In: International Conference on Learning Representations, pp. 1–35 (2018)
9. Xie, L., Lin, K., Wang, S., Wang, F., Zhou, J.: Differentially private generative adversarial network. arXiv preprint arXiv:1802.06739 (2018)
10. Wijesinghe, A., Kapoor, S., Fouda, M.M., Han, Z.: PS-FedGAN: an efficient federated learning framework based on partially shared generative adversarial networks for data privacy. arXiv preprint arXiv:2305.11437 (2023)
11. Zhang, L., Wang, X., Lu, X., Gong, M., Liu, C.: FedDPGAN: federated differentially private generative adversarial networks framework for the detection of COVID-19 pneumonia. Inf. Syst. Front. **23**(6), 1403–1415 (2021)
12. Ha, H., Hwang, U., Jang, J., Bae, H., Yoon, S.: Membership privacy-preserving GAN. In: British Machine Vision Conference (BMVC), pp. 1–13 (2022)

13. Zhang, Y., Jia, R., Pei, H., Wang, W., Li, B., Song, D.: The secret revealer: Generative model-inversion attacks against deep neural networks. In: IEEE/CVF Conference on Computer Vision and Pattern Recognition, pp. 253–261. IEEE, Los Alamitos (2020)
14. Zhang, C., Li, Z., Cao, H., Tang, B., Li, T.: Evaluating privacy leakage of generative models using trajectory analysis. IEEE Trans. Inf. Forensics Secur. **17**, 1937–1951 (2022)
15. Torkzadehmahani, R., Kairouz, P., Paten, B.: DP-CGAN: differentially private synthetic data and label generation. In: IEEE/CVF Conference on Computer Vision and Pattern Recognition Workshops, pp. 98–104. IEEE, Los Alamitos (2019)
16. Yang, R.: Privacy-preserving synthetic data generation for healthcare planning. Master's Thesis, KTH Royal Institute of Technology, Stockholm (2021)
17. Yoon, J., Drumright, L.N., Van Der Schaar, M.: Anonymization through data synthesis using generative adversarial networks (ADS-GAN). IEEE J. Biomed. Health Inform. **24**(8), 2378–2388 (2020)
18. Zhao, X., Zhang, Y., Zhang, T., Zou, D.: PEARL: data synthesis via private embeddings and adversarial reconstruction learning. In: International Conference on Machine Learning, pp. 12557–12566. PMLR (2021)

A Survey on AI-Based Anomaly Detection for Cloud Security in IIoT Environments

Jungho Kang(✉)

Baewha Women's University, Seoul, South Korea
kjh7548@naver.com

Abstract. The Industrial Internet of Things (IIoT) has revolutionized industrial operations with seamless connectivity and real-time decision-making, but the reliance on cloud computing poses significant security risks, such as data breaches and advanced persistent threats (APTs). Traditional anomaly detection methods are inadequate for the dynamic and complex IIoT-cloud ecosystem. This survey examines AI-based anomaly detection techniques, evaluating their performance in data confidentiality, real-time capabilities, scalability, and robustness against evolving cyber threats. Identified challenges include computational inefficiencies, scalability constraints, and insufficient privacy-preserving mechanisms. Based on these findings, future research directions include developing Federated Hybrid AI Frameworks that integrate Federated Learning, Reinforcement Learning, and Generative Adversarial Networks to address security and scalability challenges. Additionally, incorporating explainable AI techniques is crucial to enhance transparency and trust in detection systems. Aligning these efforts with stringent IIoT security demands is essential for building robust and adaptive frameworks for Industry 4.0.

Keywords: Cloud Security · IIoT Environment · AI-Based Anomaly Detection

1 Introduction

The Industrial Internet of Things (IIoT) has revolutionized industrial operations by enabling seamless integration of devices, systems, and processes through advanced connectivity. Industries increasingly adopt IIoT to enhance operational efficiency, scalability, and real-time decision-making. These environments often depend on cloud computing to manage and process the massive influx of data generated by connected devices. Cloud computing provides advantages such as scalability, cost efficiency, and ubiquitous access to resources, making it indispensable for IIoT applications. However, the reliance on cloud infrastructures has introduced significant security vulnerabilities, exposing IIoT environments to a range of sophisticated cyberattacks, including data breaches, distributed denial-of-service (DDoS) attacks, and malware injection [1].

The detection of anomalies, which often signals malicious activities or system faults, is a critical aspect of cloud security in IIoT. Conventional anomaly detection methods are inadequate in addressing the dynamic and heterogeneous nature of IIoT systems. The

J. S. Park et al. (Eds.): FutureTech 2025, LNEE 1483, pp. 194–201, 2025.
https://doi.org/10.1007/978-981-95-1999-6_25

inherent challenges, such as the vast scale of data, real-time operational requirements, and diverse threat vectors, necessitate advanced and adaptive solutions [2, 3]. Artificial intelligence (AI) has emerged as a transformative tool for addressing these challenges. AI-based anomaly detection techniques utilize machine learning and deep learning algorithms to identify abnormal patterns and predict potential security breaches with high accuracy and speed [4]. These methods not only improve detection capabilities but also provide scalable solutions for the complex IIoT-cloud ecosystem.

This survey focuses on the application of AI-based techniques for anomaly detection in cloud-enabled IIoT environments. The study systematically reviews recent advancements, exploring their efficacy, challenges, and practical considerations. Moreover, it emphasizes the importance of addressing unique IIoT security demands, such as safeguarding industrial data, ensuring operational continuity, and mitigating the risks associated with evolving cyber threats [5]. The scope includes an in-depth analysis of traditional and modern AI methods, identification of existing solutions, and a discussion of open issues and future research directions. By bridging gaps in research and practice, this survey aims to contribute to the development of robust security frameworks for IIoT systems, critical to the success of Industry 4.0 and beyond.

2 Related Work

2.1 Cloud Security in IIoT

The integration of IIoT with cloud computing has transformed industrial operations by enabling scalable and real-time data management but has also introduced significant security challenges, such as data breaches, DDoS attacks, and advanced persistent threats (APTs). These environments must address unique risks, including insider threats, malware injections, and ransomware, which can jeopardize system integrity and functionality. Below, we delve into each threat in detail.

- **Data Breaches:** Exposing sensitive IIoT data due to weak authentication or misconfigurations poses severe risks. Mitigation includes encryption, fine-grained access controls, AI-driven anomaly detection, and regular audits [6].
- **Distributed Denial-of-Service (DDoS) Attacks:** DDoS attacks overwhelm cloud servers with excessive traffic, disrupting critical IIoT operations and halting production. Preventative measures include scalable architectures, DDoS mitigation services, AI-based traffic analysis, and proactive network monitoring to ensure service continuity [7].
- **Malware Injection:** Malicious code compromises IIoT devices, causing disruptions and breaches. Countermeasures involve robust application security, regular updates, code verification, and runtime threat detection [8].
- **Man-in-the-Middle (MitM) Attacks:** MitM attacks compromise IIoT communications by intercepting and altering data flows, jeopardizing decision-making. Protection requires strong encryption (e.g., TLS 1.3), mutual authentication, protocol updates, and monitoring unusual activity [9].

- **Ransomware:** Ransomware encrypts critical IIoT data and systems, demanding payment and causing widespread disruptions. Defense strategies include robust backups, endpoint protection, network segmentation, and training employees to avoid phishing attempts [10].
- **Advanced Persistent Threats (APTs):** APTs are prolonged, targeted attacks using techniques like phishing and zero-day exploits to infiltrate IIoT systems for espionage or sabotage. Multi-layered security, threat intelligence collaboration, and AI-based threat detection are essential for defense [11].

To counter these evolving threats effectively, robust, mul-ti-layered defense mechanisms and AI-driven solutions are essential for ensuring data protection and operational continuity in cloud-enabled IIoT systems.

2.2 AI Techniques for Cloud in IIoT

This chapter explores AI techniques tailored for enhancing cloud-based IIoT systems. These techniques focus on improving data confidentiality, real-time processing, and overall efficiency in IIoT environments.

- **Convolutional Neural Networks (CNNs):** CNNs process structured data to detect anomalies in IIoT, analyzing network traffic and visual inputs. They enhance **Data Confidentiality** by identifying suspicious patterns and improve **Efficiency** through automated anomaly detection [12].
- **Recurrent Neural Networks (RNNs):** RNNs are well-suited for processing sequential data by retaining memory of prior inputs, enabling effective analysis of time-series patterns. In IIoT, RNNs are employed to detect behavioral anomalies and irregularities in real-time data flows, enhancing **Real-Time Capabilities** through sequence-based predictions and improving **Data Confidentiality** via robust anomaly detection [13].
- **Long Short-Term Memory (LSTM):** LSTM, an advanced variant of RNN, addresses vanishing gradient problems, enabling the modeling of long-term dependencies in data. In IIoT systems, LSTM is effective for detecting complex temporal patterns and operational trends. It enhances **Real-Time Capabilities** by capturing temporal dependencies and strengthens **Data Confidentiality** through anomaly detection in extended data sequences [14].
- **Federated Learning (FL):** FL enables decentralized AI training on local devices, preserving privacy and security by avoiding data transfers to central servers. It enhances **Data Confidentiality** and improves **Efficiency** by reducing reliance on centralized computing [15].
- **Reinforcement Learning (RL)**: RL employs agents to optimize strategies in dynamic environments using rewards and penalties. In IIoT, it enhances **Efficiency** through resource optimization and improves **Real-Time Capabilities** by enabling adaptive decision-making in complex scenarios [16].
- **Generative Adversarial Networks (GANs):** GANs utilize a generator and discriminator to create realistic data for training, anomaly detection, and network traffic reconstruction. In IIoT, GANs enhance Data Confidentiality by identifying malicious anomalies and improve Efficiency by augmenting datasets and strengthening model robustness [17].

2.3 Key Consideration

Industrial Internet of Things (IIoT) systems integrated with cloud computing face unique security challenges due to their interconnected and dynamic nature. Protecting these environments requires addressing critical considerations such as data confidentiality, efficiency, and real-time capabilities. These factors are pivotal in ensuring the seamless operation and security of IIoT systems.

- **Data Confidentiality:** In cloud-enabled IIoT systems, protecting sensitive industrial data—such as operational metrics and control commands—is critical, as unauthorized access risks increase with cloud reliance. Privacy-preserving AI techniques, such as federated learning, maintain confidentiality by keeping data local [18]. However, achieving a balance between confidentiality and usability remains a challenge, as overly restrictive security measures can limit effective anomaly detection [19].
- **Efficiency:** Limited computational resources in cloud-enabled IIoT systems require anomaly detection frameworks to optimize resource use while maintaining accuracy. Lightweight AI models, edge computing architectures, and techniques like model compression ensure efficient operations on constrained devices, reducing reliance on centralized cloud infrastructure [20].
- **Real-Time Capabilities:** Timely anomaly detection is vital for maintaining operational continuity in IIoT cloud environments. Deploying AI models at the edge allows for immediate processing and response, reducing latency [21]. Additionally, stream processing algorithms facilitate continuous monitoring and rapid threat detection, improving the overall resilience of IIoT systems.

In conclusion, addressing these key considerations ensures that AI-based anomaly detection frameworks for IIoT cloud security are robust, adaptive, and scalable. By protecting data confidentiality, optimizing efficiency, and enabling real-time operations, these frameworks can effectively address the complex security demands of IIoT environments, paving the way for safer and more reliable industrial systems.

2.4 Existing Solutions

Various studies have explored AI-based anomaly detection solutions for enhancing IIoT cloud security. This chapter examines the key techniques, achievements, and limitations of these solutions, as summarized in Table 1.

V. Priya et al. propose an ensemble classifier combining SVM, Naive Bayes, and ANN for attack detection in IIoT cloud environments, achieving 96.87% accuracy and 94.53% sensitivity. The model ensures low FPR (2.34%) and real-time scalability, but lacks detailed resource utilization analysis [22]. Song et al. introduce FDA3, a federated learning-based mechanism defending against adversarial attacks, with 98.12% accuracy and 96.45% defense success. It ensures data confidentiality through model updates but may require enhanced scalability validation [23]. Iram Bibi et al. propose the ConvLSTM2D model for IIoT threat detection, achieving 99.12% accuracy and 1.87% FPR. While effective for real-time detection, it lacks privacy-preserving measures [24]. Zolanvari et al. present an XAI framework for IIoT security, achieving 97.8% accuracy

Table 1. Summary of Existing AI-Based Anomaly Detection Solutions for IIoT Cloud Security

Author	Techniques	Objectives	Data Confidentiality	Efficiency	Real-Time
V. Priya, et al. [22]	SVM, Navie Bayes, ANN	DDoS, Data Breach	△	O	O
Song, Yunfei, et, al. [23]	Federated Learning	Adversarial Attacks	O	O	△
Iram Bibi, et al. [24]	LSTM, CNN	DDoS, Data Tampering, Unauthorized Access	△	O	O
Zolanvari, Maede, et al. [25]	CNN, XAI	DDoS, Data Exfiltration, Adversarial Attacks	△	O	△
Shah, Syed Nawaz Ali, et al. [26]	CNN, RNN	APTs,Data Manipulation, Adversarial Attacks	X	O	△
Poorazad et al. [27]	Blockchain, CNN	Intrusion Detection	O	△	△
Dhieb, Najmeddine, et al. [28]	Blockchain, Machine Learning	Adversarial Attacks	O	△	△
Safronov, Vadim, et al. [29]	AI-based Traffic Filtering	Local IoT Threat Detection	O	△	O

and 2.5% FPR. It enhances transparency but needs mechanisms for data confidentiality and real-time validation [25]. Shah et al. propose a CNN-RNN hybrid model for spatiotemporal threat detection, achieving 98.85% accuracy and 97.64% sensitivity. It ensures low FPR (1.92%) but lacks resource utilization details [26]. Poorazad et al. integrate blockchain and deep learning in IDS for IIoT, ensuring data integrity and effective anomaly detection. However, implementation complexity may limit adoption [27]. Dhieb et al. propose a blockchain-based IoT framework for security and scalability, effectively supporting large-scale deployments. Resource optimization details are minimal [28]. Safronov et al. introduce "SunBlock," a cloudless IoT protection system using AI, improving privacy and responsiveness. It efficiently mitigates threats but lacks broader threat coverage validation [29].

3 Open Issues, Solutions and Future Directions

Existing solutions for **cloud security** in IIoT environments have made notable progress but still present significant limitations. Many approaches, including ConvLSTM2D and CNN-RNN hybrids, lack robust mechanisms for **data confidentiality**, leaving sensitive data vulnerable to unauthorized access during analysis. While models like XAI and NIDS are efficient, their **real-time detection capabilities** in cloud environments often remain unvalidated, particularly under high data throughput scenarios. Blockchain-based systems, while effective for data integrity, face **scalability challenges** due to the computational overhead and complexity of integrating them into cloud platforms. Advanced AI models, such as LSTMs and GANs, also pose efficiency challenges, as their computational demands may exceed the capacity of typical cloud-hosted IIoT setups. Furthermore, many systems struggle with **robustness**, requiring frequent retraining to adapt to new and sophisticated threats like APTs and adversarial attacks.

To address these challenges, future research could focus on the development of a **Federated Hybrid AI Framework** designed specifically for **cloud security** in IIoT environments. By leveraging **Federated Learning (FL)**, the framework could enhance **data confidentiality** by processing sensitive IIoT data locally at edge devices and sharing only model updates with the cloud. This decentralized approach reduces the risks associated with data breaches and insider threats. **Reinforcement Learning (RL)** could be used to optimize **real-time detection** by enabling adaptive and autonomous decision-making in cloud-based systems, thereby reducing latency and improving response times. Additionally, **GANs** could strengthen the framework's **robustness** by augmenting training datasets with synthetic samples and detecting adversarial anomalies in cloud-hosted IIoT environments.

To ensure **efficiency**, lightweight AI models optimized through compression and pruning could be tailored for deployment in cloud systems, effectively balancing performance with resource consumption. The modular architecture of FL would also allow the framework to **scale** seamlessly across cloud infrastructures, accommodating the diverse and growing needs of industrial applications. Finally, integrating **XAI techniques** would provide transparency into the anomaly detection logic, empowering cloud operators to make informed decisions and build trust in the system's predictions.

Exploring such a research direction could pave the way for advanced, adaptive, and scalable solutions that directly address the unique challenges of securing IIoT cloud infrastructures, ultimately enabling safer and more efficient industrial systems.

4 Conclusion

The integration of cloud computing with IIoT has revolutionized industrial operations by offering scalability, cost efficiency, and real-time data management. However, these advancements have introduced significant security challenges, making anomaly detection a critical aspect of safeguarding cloud-enabled IIoT systems. This study highlights the importance of leveraging AI-based techniques to address these challenges, emphasizing key requirements such as data confidentiality, efficiency, real-time capabilities, scalability, and robustness against evolving threats. By systematically reviewing existing solutions, the study identifies gaps in current approaches, including limitations in ensuring data privacy, handling computational demands, and adapting to sophisticated threats such as advanced persistent threats. These directions emphasize the importance of developing decentralized, scalable, and explainable solutions capable of meeting the unique demands of IIoT cloud infrastructures. Advancing these research initiatives will play a critical role in enabling secure and efficient industrial systems.

Acknowledgment. This work was financially supported by Baewha Women's University.

References

1. Douiba, M., et al.: An improved anomaly detection model for IoT security using decision tree and gradient boosting. J. Supercomput. **79**(3), 3392–3411 (2023)
2. Kemnitz, J., et al.: Towards robust and transferable IIoT sensor based anomaly classification using artificial intelligence. In: International Data Science Conference. Springer Fachmedien Wiesbaden, Wiesbaden (2021)
3. Jadidi, Z., et al.: Security of machine learning-based anomaly detection in cyber physical systems. In: 2022 International Conference on Computer Communications and Networks (ICCCN). IEEE (2022)
4. DeMedeiros, K., Hendawi, A., Alvarez, M.: A survey of AI-based anomaly detection in IoT and sensor networks. Sensors **23**(3), 1352 (2023)
5. Zolanvari, M., Ghubaish, A., Jain, R.: Addai: anomaly detection using distributed AI. In: 2021 IEEE International Conference on Networking, Sensing and Control (ICNSC), vol. 1. IEEE (2021)
6. Singh, N., Buyya, R., Kim, H.: Securing Cloud-Based Internet of Things: Challenges and Mitigations. arXiv preprint arXiv:2402.00356 (2024)
7. Berchi, R., Louail, L., Cherbal, S.: Security issues in cloud-based IoT systems. In: 2023 5th International Conference on Pattern Analysis and Intelligent Systems (PAIS). IEEE (2023)
8. Alnajim, A.M., et al.: A comprehensive survey of cybersecurity threats, attacks, and effective countermeasures in Industrial Internet of things. Technologies **11**(6), 161 (2023)
9. Cecílio, J., Souto, A.: Security issues in industrial internet-of-things: threats, attacks and solutions. In: 2024 IEEE International Workshop on Metrology for Industry 4.0 & IoT (MetroInd4. 0 & IoT). IEEE (2024)
10. Mehdipour, F.: A review of IoT security challenges and solutions. In: 2020 8th International Japan-Africa Conference on Electronics, Communications, and Computations (JAC-ECC). IEEE (2020)
11. Zolfaghari, B., et al.: The dichotomy of cloud and IoT: Cloud-assisted IoT from a security perspective. arXiv preprint arXiv:2207.01590 (2022)
12. Yan, W.Q.: Convolutional neural networks and recurrent neural networks. In: Computational Methods for Deep Learning: Theory, Algorithms, and Implementations, pp. 69–124. Springer Nature Singapore, Singapore (2023)
13. Zucchet, N., Orvieto, A.: Recurrent neural networks: vanishing and exploding gradients are not the end of the story. arXiv preprint arXiv:2405.21064 (2024)
14. Tyagi, G., Ray, S.: A comparative review of convolutional neural networks, long short-term memory, and recurrent neural networks in recommendation systems. In: International Conference on Artificial Intelligence on Textile and Apparel. Springer Nature Singapore, Singapore (2023)
15. Li, T., et al.: Federated learning: challenges, methods, and future directions. IEEE Sig. Process. Mag. **37**(3), 50–60 (2020)

16. Wang, X., et al.: Deep reinforcement learning: a survey. IEEE Trans. Neural Netw. Learn. Sys. **35**(4), 5064–5078 (2022)
17. Wang, Z., She, Q., Ward, T.E.: Generative adversarial networks: a survey and taxonomy. arXiv preprint arXiv:1906.01529 2 (2019)
18. Bin Mofidul, R., et al.: Real-time energy data acquisition, anomaly detection, and monitoring system: Implementation of a secured, robust, and integrated global IIoT infrastructure with edge and cloud AI. Sensors **22**(22), 8980 (2022)
19. Arazzi, M., Nicolazzo, S., Nocera, A.: A fully privacy-preserving solution for anomaly detection in IoT using federated learning and homomorphic encryption. Inform. Syst. Front. 1–24 (2023)
20. Konatham, B., et al.: A Secure Hybrid Deep Learning Technique for Anomaly Detection in IIoT Edge Computing. Authorea Preprints (2024)
21. Raeiszadeh, M., et al.: Real-time adaptive anomaly detection in industrial IoT environments. IEEE Trans. Netw. Serv. Manag. (2024)
22. Priya, V., Thaseen, I.S., Gadekallu, T.R., Aboudaif, M.K., Nasr, E.A.: Robust attack detection approach for IIoT using ensemble classifier. Comput. Mater. Contin. **66**(3), 2457–2470 (2021)
23. Song, Y., et al.: FDA3: federated defense against adversarial attacks for cloud-based IIoT applications. IEEE Trans. Indust. Inform. **17**(11), 7830–7838 (2020)
24. Bibi, I., Akhunzada, A., Kumar, N.: Deep AI-powered cyber threat analysis in IIoT. IEEE Internet Things J. **10**(9), 7749–7760 (2022)
25. Zolanvari, M. et al.: Trust xai: model-agnostic explanations for AI with a case study on IIoT security. IEEE Internet Things J. **10**(4), 2967–2978 (2021)
26. Shah, S.N.A., et al.: A novel deep learning-based approach to encounter cyber threats in IIoT. In: 2022 Mohammad Ali Jinnah University International Conference on Computing (MAJICC). IEEE (2022)
27. Poorazad, S.K., Benzaïd, C., Taleb, T.: Blockchain and deep learning-based IDS for securing SDN-enabled industrial IoT environments. In: GLOBECOM 2023–2023 IEEE Global Communications Conference. IEEE (2023)
28. Dhieb, N., et al.: Scalable and secure architecture for distributed IoT Systems. In: 2020 IEEE Technology and Engineering Management Conference (TEMSCON). IEEE (2020)
29. Safronov, V., et al.: SunBlock: cloudless protection for IoT systems. In: International Conference on Passive and Active Network Measurement. Springer Nature Switzerland, Cham (2024)

Author Index

J. S. Park et al. (Eds.): FutureTech 2025, LNEE 1483, pp. 203–204, 2025.
https://doi.org/10.1007/978-981-95-1999-6

MIX
Papier aus verantwortungsvollen Quellen
Paper from responsible sources
FSC® C105338

If you have any concerns about our products,
you can contact us on
ProductSafety@springernature.com

In case Publisher is established outside the EU,
the EU authorized representative is:
Springer Nature Customer Service Center GmbH
Europaplatz 3, 69115 Heidelberg, Germany

Printed by Libri Plureos GmbH
in Hamburg, Germany